Seaweeds

David N. Thomas

SMITHSONIAN INSTITUTION PRESS, WASHINGTON, D.C.
IN ASSOCIATION WITH THE NATURAL HISTORY MUSEUM, LONDON

To George. Thank you.

Published in the United States of America
by the Smithsonian Institution Press in association with
The Natural History Museum, London
Cromwell Road
London SW7 5BD
United Kingdom

Library of Congress Cataloging-in-Publication Data
Thomas, David N. (David Neville), 1962–
      Seaweeds / David N. Thomas.
            p.   cm.
      Includes bibliographical references.
      ISBN 1-58834-050-3 (alk. paper)
            1. Marine algae. 2. Marine algae—Ecology. I. Title.
      QK570.2.T46   2002
      579.8—dc21   2002023141

Manufactured in Singapore, not at government expense
09 08 07 06 05 04 03 02 5 4 3 2 1

Edited by Celia Coyne
Designed by Mercer Design
Reproduction and printing by Craft Print, Singapore

On the front cover: *Fucus vesiculosus*
On the back cover and title page: *Saccorhiza polyschides*

Contents

Preface

This book looks at the seaweeds that inhabit the transition zone between the ocean and land, concentrating on mainly fully marine species, big enough to be clearly seen. Although they photosynthesise, seaweeds are algae and differ from plants growing on land in that they do not flower and have no roots, leaves or highly organised tissues for transporting water and nutrients.

Seaweeds are found in seawater and brackish waters across the globe. They range in size from a few centimetres up to 50 m (164 ft) giants. Some are permanently submerged and form dense underwater forests that fringe the coasts. Others are exposed to the air during low tides where they can be shrivelled to a crisp in the baking summer sun or frozen solid on a cold winter's day. As well as sudden changes in temperature, seaweeds have to cope with the pounding of waves and strong coastal currents, while competing for light, nutrients and space. In spite of such harsh living conditions, seaweeds are ubiquitous and have long been a source of food for humans and animals. They are also used in fertilisers, medicine and in the production of cosmetics. Coastal developments, pollution and the introduction of non-native seaweed species threaten natural seaweed populations; conserving this valuable resource continues to be a major challenge.

These issues are discussed here following a wide-ranging introduction to how seaweeds survive the rigours of life in the ocean.

Author

David Thomas is a senior lecturer in biological oceanography at the University of Wales, Bangor. He has worked with seaweed since his final undergraduate project 20 years ago, which was concerned with the seaweeds on the shores of the Irish Sea. Since then he has investigated the physiology and ecology of seaweeds in the Antarctic, Arctic, Baltic and South China Seas. He was awarded his PhD from the Botany Department, University of Liverpool in 1988. Before taking up his position in Bangor in 1996, he held four research positions in Germany at the Universities of Bremen and Oldenburg, the Alfred Wegener Institute for Marine and Polar Research and the Centre for Marine Tropical Ecology. In addition to seaweeds his research interests are concerned with the plankton of polar regions with a particular emphasis on seasonally ice-covered waters.

What are seaweeds?

"It was once prettily said by a lady who cultivated flowers, that she had 'buried many a care in her garden'; and the sea-weed collector can often say the same of his garden – the shore; as many a loving disciple could testify, who, having taken up the pursuit originally as a resource against weariness, or a light possible occupation during hours of sickness, has ended by an enthusiastic love, which throws a charm over every sea-place on the coast…"
(Mrs Gatty, *British Seaweeds*, 1872).

Mrs Gatty was, without a doubt, a seaweed enthusiast. Maybe her words appear a little over the top, but the seaweeds that cover our rocky shorelines and wash up on our beaches after a storm are a diverse group with a myriad shapes and colours. However, seaweeds are not universally appreciated and the emotions they evoke have even been known to divide communities. In 1997 the Cornish village Porthcothan was reported in the British press as "going to war over the fate of rotting seaweed". One faction insisted

LEFT **A falling tide exposes *Fucus serratus* and other low shore seaweeds.**

that the seaweed was vital for the ecology of the beach, providing refuge for sand-hoppers, food for birds and fish, and even anchoring the local sand dunes. The opposition wanted to remove the unsightly, smelly weed and maintain a pristine beach to attract tourists.

Whether seaweeds are more of a nuisance than a delight is clearly open to debate.

BELOW **Large brown seaweeds can form dense underwater forests, as here at Easter Islands, part of the Scilly Isles.**

However, mankind has used them for at least the past 2500 years, as ancient Chinese documents testify. Phycologists (from *phykos*, the Greek word for seaweed) have been studying seaweed for hundreds of years, yet we are still only scratching at the surface of what there is to be known about their evolution, ecology and physiology.

Are they weeds?

An easy definition of seaweeds is an almost impossible task, at least in a way that doesn't demand extensive botanical training to help grapple with subtleties that have occupied taxonomists for hundreds of years. From the outset we should drop the idea of a 'weed' as an appropriate description. Weed seems to infer that seaweeds are fast-growing nuisances that are difficult to get rid of. Although seaweeds can be both of these things, they are more appropriately considered as a group containing the aquatic equivalent of trees, shrubs, bushes and lawns. The different types of seaweed offer a rich variety of habitats and refuges for animals, as well as being an important source of food.

Even the 'sea' part of seaweed is somewhat misleading since several species are highly tolerant to changes in the salt concentration of the water in which they grow, and can be found in both marine and brackish environments such as estuaries. They can even be found in land-locked freshwaters, often a long way away from the sea.

Seaweeds are algae

Seaweeds are classified as algae – photosynthetic organisms which, in contrast

to the plants growing on land, are non-flowering and do not have roots, leafy shoots or sophisticated tissues for transporting water, sugars and nutrients. The taxonomic descriptions and the relationships of major evolutionary lineages remain controversial and are subject to ongoing research. The fossil record shows that there were seaweeds (with a similar appearance to the species we see today) living at least 500 million years ago. However, such fossils are rare, since the soft tissues of many of the seaweeds tend not to preserve very well.

The seaweeds have several plant-like traits and for many years their biology was taught together with that of the fungi in 'lower plant' biology classes that reflected the relative lack of organisation and tissue differentiation when compared with mosses, ferns and flowering plants. However, besides the green species, recent advances in molecular biology show that most seaweeds are only remotely related to land plants.

Some of the taxonomic frustration for the non-specialist lies in the fact that algae range in size from microscopic single cells that grow in the plankton found in oceans, lakes and rivers through to giant seaweeds over 50 m (164 ft) long that form dense forests in coastal waters. It is such a diverse group that in many ways the connections between them are not so very obvious. Algae don't even have to live in water. For example, many of the green stains you see on the bark of trees or on the side of buildings are the result of dense growths of microscopic algae.

From encrustations to giants

Seaweeds are therefore not plants, but rather photosynthetic aquatic organisms. They are often referred to as macroalgae (large algae) and for many species, such as the giant kelps, it is obvious why. However, there are plenty of seaweeds that are very small indeed and only visible by using a good magnifying lens. In some cases the form and structures that are

ABOVE LEFT **Seaweeds produce oxygen during photosynthesis, seen here collecting as bubbles inside the tube-like thallus of an** *Enteromorpha* **species.**

ABOVE RIGHT **A variety of red, green and brown seaweeds at low tide.**

RIGHT **Seaweeds not only grow on rocks, but also on virtually any solid structure covered periodically by the sea.**

important for telling the various species apart can only be seen with a high powered microscope.

Macroalgae are, for the most part, multicellular organisms. Some are hardly more than green, red or pink encrustations on the surfaces of rocks that look as though they should be grouped together with the lichens. Others are huge and look as though they have come from a fantasy filmset or some prehistoric time.

Where do seaweeds grow?

Macroalgae need water to grow in, although it is evident from the multitude of seaweed species that are exposed on a rocky shore every time the tide goes out that many do not need to be continually submerged. The

seaweeds occupy a region that is in transition from the aquatic to the terrestrial way of life. While many species are fully adapted to the aquatic existence and are damaged even after being out of the water for a short time, there are other species higher on the shore that can remain 'high and dry' for long periods.

Generally seaweeds grow attached to surfaces that are bathed by water. These can be the underside of boats, ropes, piers, boulders, rock faces – in fact, any surface can form an anchorage for seaweeds. Before the advent of anti-fouling paints, getting rid of seaweeds on marine structures was a major industry. Until recently, navies employed phycologists to research ways of ridding ships' hulls of dense algal growths, a problem commented on by Plutarch (AD 47–127) who wrote that it was usual to "scrape the weeds, ooze and filth from the ships' sides to make them go more easily through the water".

Some seaweeds form dense turfs on sandy or muddy expanses of shore, with the appearance of underwater lawns, while others attach themselves to the shells of crabs, or grow on the surfaces of other seaweeds. The decorator crabs, a group including many different crab species, adorn themselves with seaweeds alongside other materials from their surroundings as a highly effective form of camouflage.

Though many species grow attached, seaweeds can also be free-floating, such as the *Sargassum* of the Sargasso Sea, which can drift on ocean currents to occupy areas covering thousands of square kilometres. Masses of floating seaweed can become entrapped in bays or develop in shallow coastal waters such as the dense accumulations of green algae that have periodically clogged up parts of the Baltic and Adriatic seas.

Seaweeds are red, green and brown

From 'mere' encrustations to giant kelps as tall as giant redwood trees, all seaweeds need light and carbon dioxide for photosynthesis, and sources of nitrogen, phosphorus and

LEFT **Part of the convoluted stipe of the brown seaweed *Saccorhiza polyschides*.**

RIGHT **The green seaweed *Codium* and brown *Leathesia difformis* growing among young fronds of red, brown and calcified seaweeds.**

other trace elements for growth. Just like any other photosynthetic organism, they produce oxygen as a result of their photosynthesis during the day and carbon dioxide during respiration, which goes on both day and night. The basis of photosynthesis is the light-harvesting pigment chlorophyll a, which is synthesised and stored in the chloroplasts within the cells. This gives seaweeds their greenish colour. Seaweeds also contain a range of other pigments that are used to capture light for photosynthesis or to protect against harmful ultraviolet radiation. The pigments present in a particular species of seaweed are used as a means of classification: red seaweeds belong to the division of Rhodophyta, green seaweeds belong to the Chlorophyta division, and brown seaweeds belong to the class Phaeophyceae in the division Heterokontophyta.

Brown seaweeds are brown because they have high concentrations of carotenoids such as yellowish-brown fucoxanthin. The red seaweeds are red because they contain red and even blue pigments such as phycoerythrin, phycocyanin and allophycocyanin that combine in various ways to mask the green of the chlorophyll a within the cells. Green seaweeds also contain accessory pigments, but these do not screen the dominating green of the chlorophyll a. However, as countless generations of seaweed enthusiasts will confirm, colour is an awkward diagnostic tool since the hues and tones are so variable: brown seaweeds may look more like an olive green, whereas the difference between a deep burgundy and a dark brown can be difficult to discern at times.

Numbers of seaweed species

Taxonomists who specialise in the identification, naming and classification of all algae (micro- and macroalgae) debate fiercely about the numbers of species that exist. Some estimate there to be around 36,000-50,000 species, although values of up to 10 million are quoted. Looking just at the seaweeds, numbers vary significantly between the three major groupings. Worldwide there are about 1500-2000 brown seaweed species, 5000-6000 species of red seaweed and about 1000-2000 species of green seaweed. The brown and red seaweeds are predominantly marine species, whereas a much higher percentage of green seaweeds also grow in brackish and freshwaters.

Considering the range of aquatic habitats in which they are found a compromise has to be reached for the macroalgae to be discussed here. It seems pertinent to focus on marine species. Discussion will also be mostly restricted to the larger species of macroalgae that are likely to be encountered while walking on the beach, snorkelling or diving.

What do seaweeds look like?

It was probably the beauty of seaweeds that spurred on the 18th and 19th century pioneer 'seaweed hunters', together with a thirst for cataloguing the weird and unusual. The tapestry of colour seen in a rockpool, where splashes of encrusting pink are set against a backcloth of vivid greens, deep burgundies and browns of every hue, is enough to inspire any naturalist.

Seaweed form: the weird and the wonderful

Even a cursory look at a selection of seaweeds reveals that for many species there is no obvious differentiation in form: one part of the thallus (a term used to describe the whole of a seaweed) is very much like the rest. However, other species have a more plant-like appearance with a structure that anchors

BELOW LEFT **Rock pools can harbour a rich diversity of seaweed species to form a rich tapestry of colour and form.**

OPPOSITE PAGE

TOP LEFT **The tropical green seaweed *Rhipocephalus phoenix*.**

MIDDLE LEFT *Acetabularia acetabolum* **looks like a clump of inverted umbrellas growing alongside mounds of dark green *Codium bursa*.**

BOTTOM LEFT **A sheet of *Ulva*, the sea lettuce.**

TOP RIGHT **Chains of *Chaetomorpha coliformis*. Each bead is a single cell.**

MIDDLE RIGHT **The brown species *Ascophyllum nodosum* with well developed air bladders.**

BOTTOM RIGHT **The small button-like structures are the brown seaweed *Himanthalia elongata* from which the long rope-like reproductive structures are produced.**

RIGHT **A young kelp attached to an oyster shell by a holdfast. The fronds are supported on a stalk-like stipe.**

RIGHT **A young kelp attached to an oyster shell by a holdfast. The fronds are supported on a stalk-like stipe.**

them (holdfast), a stem or stalk (stipe) and leaf-like blades (fronds) coming off the stipe.

Some species have calcified tissues that are hard and brittle: the red *Corallina*'s ribbons of segmented boxes feel like a collection of small stones when you pick them up. The brown *Hormosira banksii* drapes rocks in cascades of hollow, gas-filled beads, whereas the green *Udotea* and brown *Padina* form flattened fans that are characteristic of seagrass meadows and coral reefs. *Halimeda* is a rugged green seaweed consisting of flattened, calcified segments interspersed with flexible joints. It is a stark contrast to the delicate green *Acetabularia*, the mermaid's wine glass, that grows in clumps of miniature inverted umbrellas. Others, including the green *Ulva* and red *Porphyra*, are fragile sheets only a

few cells thick. Green *Enteromorpha* species and the red *Halosaccion glandiforme* are a little more sophisticated in that the sheets have formed into tubes or sacks that fill up with seawater to form rigid structures.

Brown *Ectocarpus* and green *Cladophora* are really no more than fine filaments of single cells, whereas in many of the fine red species and the green *Codium* the filaments are fused and organised into highly complex structures. The most dramatic filaments are those of the emerald green *Chaetomorpha colifomis*, which forms chains of transparent bead-like cells, each cell up to 8 mm (0.4 in) in diameter. But these chains are minuscule when compared to the tropical single-celled *Ventricaria ventricosa*. Its dark green iridescent spheres can be up to 3 cm (1.2 in) across and look like huge glass marbles attached to the rocks.

There are a whole host of large, thick, membranous seaweeds, such as the brown *Fucus* and *Ascophyllum*, which are generally shorter than a metre, regularly branched and often studded with bladders. Some, such as the brown *Postelsia*, resemble miniature palm trees, their thick stems crowned with a tuft of leaf-like blades. Others, such as the brown *Chorda*, are whip-like single blades that resemble ropes as they lie in twisted masses on temperate shores.

The seaweed giants

For many people it is the larger brown species, often referred to as 'kelps', that best conjure up the image of seaweed. These seaweeds, including *Alaria*, *Laminaria*, *Ecklonia* and *Durvillaea*, are several metres long with large leathery fronds held on slender, flexible stalks that emerge from the water only on the lowest spring tides. The real giants of the seaweed world, such as *Macrocystis pyrifera* and *Nereocystis luetkeana*, grow as fast as half a metre a day to tower 30–50 m (98–164 ft) above the seabed. These are the canopy species of underwater forests with long slender stems supporting a crown of blades near the water surface.

We most often encounter seaweeds when they are draped over rocks, often in large tangled masses. But if we get under the water we can fully appreciate the form of seaweeds. Only when they are surrounded by water can we really see why they are shaped the way they are and appreciate the structural adaptations that enable them to function.

Holding fast

Seaweeds do not have the root systems we are familiar with from terrestrial plants. Roots are not needed since nutrients such as nitrogen, phosphorus and trace elements are dissolved in the seawater. The nutrients can be taken up and exchanged by diffusion and active transport directly across all the surfaces of the seaweed (carbon dioxide and oxygen are also assimilated and released in this way). Therefore it is only the mechanical features of a root system that would be beneficial for the seaweeds, holding them steady no matter how turbulent the water movement is.

Many types of seaweed have a very obvious holdfast that on first inspection looks

BELOW **The massive brown seaweed *Durvillaea* attaches itself to rocks with a huge disc-shaped holdfast.**

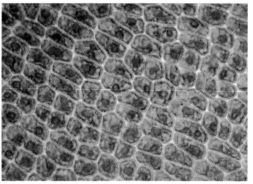

FAR LEFT *Caulerpa* fronds are securely anchored into sandy beds by horizontally spreading rhizomes that also help stabilise the sediment.

LEFT Seaweeds carry green chlorophyll in chloroplasts inside the cells, as in these *Enteromorpha* cells (x100).

like a jumbled mass of roots. The holdfast is made up of branched, thickened tissues known as haptera that attach the thalli to rocks and boulders, stones or even molluscs and crabs with strong adhesives and fine hairs. In some cases, such as in the brown *Durvillaea*, the holdfast is a disk rather than a root-like mass. The size of the holdfast decreases with the size of the seaweed: a large *Macrocystis* may have a holdfast as big as a football, while a *Laminaria* or *Durvillaea* will have fist-sized holdfasts and a *Fucus* will have a small disc-like holdfast. Pull on any of the fronds of these seaweeds and you will soon see how strong an attachment device these holdfasts can be: nine times out of ten you'll rip the fronds from the holdfast which will stay put on the rocks.

Rhizoid holdfasts

In sandy or silty seabeds species of green seaweed such as *Halimeda, Udotea* and *Penicillus* develop holdfasts that are made from delicate root-like rhizoids. These structures serve to minimalise disturbance and uprooting caused by animals grazing the surface sediments or burrowing into them. In

this case the holdfast is acting more like the roots of a terrestrial plant, stabilising the seaweed in a loosely aggregated soil, rather than holding it tight to a largely impenetrable substratum. Species of *Caulerpa*, also commonly found in sandy beds and lagoonal environments, form extensive mats of horizontally growing root-like structures – similar to the rhizomes of some terrestrial plants – which are anchored by rhizoids.

Acid attack

Some seaweeds, such as the conchocelis phase of *Porphyra* (see p. 25), obtain a foothold using a more aggressive approach. They secrete acid which dissolves the calcium carbonate of shells and coral skeletons so that fine rhizoids and filaments can grow through the bored out matrix. The calcium and bicarbonate that are released by the acidification of the calcium carbonate structures also provide a localised source of nutrients for the seaweed.

The quest for light

Light, along with a supply of carbon dioxide and water, is the fundamental requirement for photosynthesis so seaweed must maximise the amount of light hitting the chloroplasts in its tissues to achieve maximum growth. Seaweeds have chlorophyll-containing chloroplasts in most surface tissues and therefore photosynthesis is not restricted to specialist structures (such as the leaves of terrestrial plants). In many of the seaweeds that do not have any obvious differentiation between tissues, branching patterns often enhance the way in which light hits the chloroplast-containing cells.

In seaweeds where there is some differentiation in structure, some parts of the thallus are more efficient at trapping the incident light than others; for example, the fronds of *Macrocystis* lie flat in the water to maximise the surface area exposed to the light. Clearly these trap more light than the

RIGHT **At the base of each blade of the giant brown seaweed *Macrocystis pyrifera* is a single air bladder that buoys the plants up to the surface of the sea, tens of metres above the holdfast below.**

long, relatively thin stipe supporting the leaf-like fronds.

In cloudy coastal waters full of plankton and suspended particles washed in from the land and brought down by rivers, light may not penetrate further than a few metres. There are several structural features, common to many species of seaweed, that help to address this problem.

The stipe

The stipe of a seaweed can be very long indeed. The purpose is to support the bulk of the photosynthetic tissue at the surface of the water where incident light is at its maximum intensity. *Nereocystis luetkeana* is a superb example of this, with stipes that are over 30 m (98 ft) long supporting up to 100 fronds, each several meters long. In this case the stipe is acting very much like the trunk of a tree suspending the maximum surface area of photosynthetic cells as close to the incident light as possible.

Buoyancy aids

But it isn't just the long stipe of *Nereocystis luetkeana* alone that guarantees the blades sit at the top of the water bathed in light. In fact, such a long stipe is rather too flexible and the blades only sit on the surface because the apex of the stipe terminates in a 15 cm (6 in) diameter float, or pneumatocyst. This large gas-filled bladder ensures that the blades are floating as close to the surface of the water as possible. That other giant, *Macrocystis pyrifera*, doesn't rely on one large float to support it in the water – at the base of each of its thousands of fronds is a small gas bladder connected to

17

LEFT The giant air bladder, up to 15 cm (6 in) in diameter, of the brown seaweed *Pelagophycus porra*.

LEFT The air bladders lift the fronds of *Ascophyllum nodosum* to the surface of the water, ensuring maximum exposure to the sunlight.

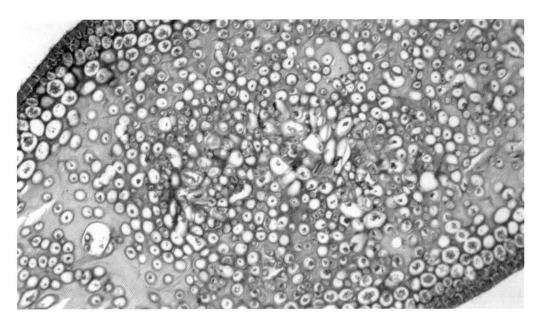

LEFT **A section through** *Fucus*, **showing the differentiation of tissue into an outer meristoderm one cell thick, a layer of larger cortical cells and a diffuse innermost medulla region.**

the highly branched stipe. Many other species use gas-filled bladders in a similar way, including *Ascophyllum*, *Sargassum*, *Fucus* and the bead-like *Hormosira*. The bladders contain oxygen and nitrogen in roughly the same proportion as air and varying amounts of carbon dioxide. Curiously the huge pneumatocysts of *Nereocystis* contain up to 10% carbon monoxide, although it is unclear why.

Organisation of tissues

In 'simple' seaweeds each cell is responsible for its own nutrition. However, several of the larger brown species have specialised tissues that enable metabolites to be transferred from one part of the seaweed to another. The transport system is not as sophisticated as that seen in terrestrial plants, but is effective nevertheless. In species of *Laminaria*, *Macrocystis*, and to a lesser degree *Fucus*, the stipes and fronds typically have an outer layer of pigmented cells (meristoderm) overlying a layer of colourless cells (cortex) with, in some species, an innermost medulla core. In the innermost layers of *Laminaria* and *Macrocystis* there are specialised elongated cells (trumpet cells). The trumpet cells have perforated end walls (sieve plates) that can be arranged end to end to form sieve tubes. Sugars and amino acids generated in the photosynthetic cells are transported through these tubes to different parts of the seaweed. This is crucial to a seaweed that is tens of metres long since it means that these chemical products of photosynthesis reach the cells positioned at depths where light levels are low. Experiments with *Macrocystis*, where the sugars produced during photosynthesis were labelled with radioactive carbon, have shown that the sugars can move at a rate of up to 50 cm (20 in) per day away from the photosynthesising fronds.

19

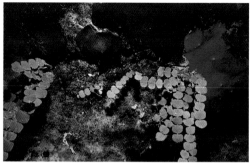

Calcareous seaweeds

A curious feature of one large group of seaweeds is that they are rather hard and brittle to touch. These calcareous seaweeds – species that have deposited calcium carbonate in the form of calcite or aragonite crystals within their tissues – can be brown, red or green. The calcification can be so extensive that it becomes an important process in cementing reefs together. The two forms of crystal never occur together in the same seaweed and there is still some debate as to the metabolic processes that actually lead to the deposition.

In some species the crystals are laid down outside of the cells, such as in the green *Halimeda* where aragonite crystals precipitate in the spaces between the cells. In the fan-shaped brown *Padina*, aragonite is sometimes precipitated in concentric bands on the outer surface of the thallus. The red coralline seaweeds, such as *Lithothamnion* and *Lithophyllum*, are so named because they closely resemble and are commonly mistaken for corals. They deposit calcite within their cell walls to the extent that the cells become encased, except for the cellular connections.

Certain polysaccharides (carbohydrates made up of many simple sugars) in seaweed cell walls can block crystal growth. Having these polysaccharides in the cell walls is the likely reason why many marine seaweeds are not calcified. After all, a thick layer of calcium carbonate effectively stops light getting to the cells below, therefore reducing photosynthesis. On the other hand a stone-like texture is not very palatable to most grazers.

ABOVE LEFT **A calcified red *Peyssonnelia squamaria* from Bluehole, Bwejra, Gozo.**

TOP RIGHT **The calcified green *Halimeda copiosa* from Grand Turk, Caribbean.**

BOTTOM RIGHT **Fan-shaped brown *Padina boergesenii* in which calcium carbonate is laid down in concentric bands; Shuwaihat, Abu Dhabi.**

Seaweed growth and reproduction

Seaweeds have complicated life cycles. In general, there is an asexual phase (sporophyte), when the seaweed's cells are diploid (containing two copies of each chromosome), followed by a sexual phase (gametophyte), when the cells are haploid (containing one copy of the chromosomes). Following fusion of male and female gametes (both haploid) the chromosome number is doubled and there has to be a mechanism for restoring the haploid number before the next generation of gametes and ensuring effective distribution and exchange of genetic material within the species. This is obtained by an alternation of generations between the sporophyte and gametophyte stages.

The idea of a strict alternation between sporophyte and gametophyte is confusing because often there is no regular pattern of progression from one phase to the next. During the asexual phase reproduction is via spores or by the fragmentation of the thallus,

LEFT **Fronds of the brown *Laminaria* revealed at low tide.**

whereby a new seaweed thallus grows from a piece that has broken off. During the sexual phase of any life cycle gametes (eggs and sperm) are produced and these have to be brought into contact with each other. This is not an insignificant task in the aquatic world, where anything small released into the water is likely to be swept swiftly away.

Asexual reproduction allows for efficient population increases of a particular species, with the disadvantage that no new genetic variation is introduced to the progeny of the parent seaweed. In contrast, sexual reproduction allows such variation to be introduced, but is more costly because of the waste of gametes that fail to fuse.

Life cycles

There is a tremendous variation on this general theme and the subject of seaweed life cycles is a minefield of exceptions, alternatives and idiosyncratic pathways. Life cycles remain unknown for many species and often when it is thought that a researcher has worked out a whole life cycle for a particular species an intriguing exception becomes apparent. On pages 22–26 the contrasting life cycles of three species are described. They have been selected because they show quite different strategies in their sporophyte and gametophyte generations and in the dispersal of progeny.

Life cycle of *Laminaria*

The large brown form of *Laminaria* (a), seen on the lower shore at extreme low waters, is the diploid sporophyte generation of the seaweed. Within specialised structures, known as unilocular sporangia (b), meiosis takes

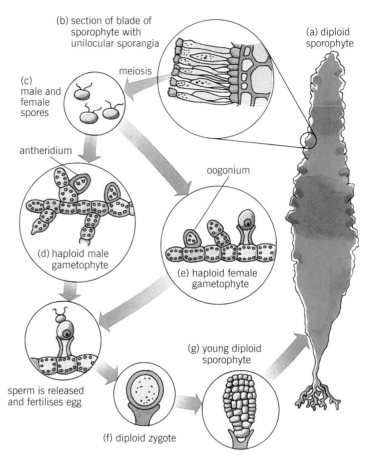

ABOVE **Life cycle of *Laminaria*.**

place to produce equal numbers of microscopic male and female spores (c) (haploid). The spores are released and settle on to the substratum, where they germinate into male (d) and female (e) gametophytes (still haploid) that are finely branched encrusting forms, only visible with a magnifying lens. The male is more branched and has smaller cells than the female. The male produces antheridia, each of which contains a single sperm, whereas the female thalli produce oogonia, each one containing a single egg. The egg is fertilised by the sperm to produce a zygote (f) (diploid) that develops into a young

RIGHT *Laminaria* gametophytes.

FAR RIGHT **Unilocular sporangia on the surface of a *Laminaria* blade.**

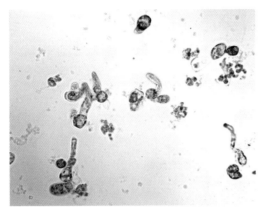

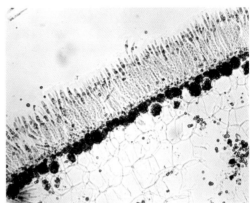

sporophyte (g). This then grows into the large thalli we are most familiar with (a).

Life cycle of *Fucus*

A large number of brown species such as *Fucus*, *Pelvetia* and *Ascophyllum* have similar life cycles of a different pattern. These seaweeds

dominate most areas of intertidal rocky shores and the thalli that we recognise are diploid sporophytes (a). When the thalli reach reproductive maturity the tips of the fronds develop into swollen, pitted structures (receptacles, b) that are often a different colour from the rest of the frond. The sunken flask-

RIGHT **Life cycle of *Fucus*.**

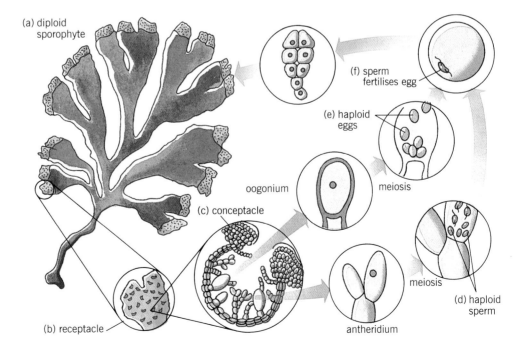

(a) diploid sporophyte

(f) sperm fertilises egg

(e) haploid eggs

meiosis

oogonium

(c) conceptacle

(d) haploid sperm

meiosis

(b) receptacle

antheridium

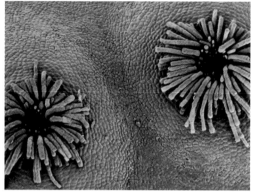

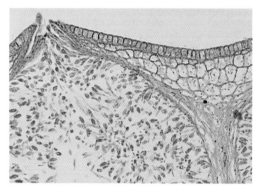

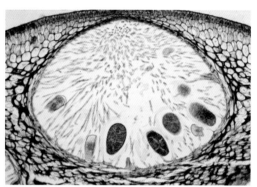

TOP LEFT **Receptacle of *Fucus spiralis*, studded with many raised conceptacles.**

TOP RIGHT **Scanning electron micrograph of two *Fucus* conceptacle openings and the hair cells.**

BOTTOM LEFT **Section through a *Fucus* male conceptacle showing the antheridia from which the sperm are produced.**

BOTTOM RIGHT **Section thorough a *Fucus* female conceptacle showing the oogonia from which the eggs are produced.**

shaped pits are called conceptacles (c) that in some monoecious species (which have both male and female structures on the same thallus), such as *Fucus spiralis*, contain oogonia, antheridia and sterile hair cells (paraphyses). In dioecious species (where the thalli are either male or female), such as *Fucus serratus*, the conceptacles on a single thallus contain either oogonia or antheridia, but not the two together. After meiosis the antheridia produce large numbers of haploid sperm (d) and the oogonia produce between one and eight haploid eggs (e). When ripe, the eggs are released and fertilised by the sperm (f) to produce diploid zygotes that develop into the sporophyte thalli.

Life cycle of *Porphyra*

The red seaweed *Porphyra* has a more obvious alternation of generations. There are about 70 *Porphyra* species worldwide, found as thin purple sheets on the low to mid shore (and sold as nori in Asian food shops or sushi restaurants). The thallus that we see, and which is commercially important, is haploid (a). It can reproduce asexually by forming spores that are released and which germinate to form replicate thalli (b). In different zones on a single seaweed both male (c) and female (d) gamete-producing cells can form. The female gametes (carpogonia), still in the thallus, are fertilised by released sperm. The now diploid carpogonia divide to produce

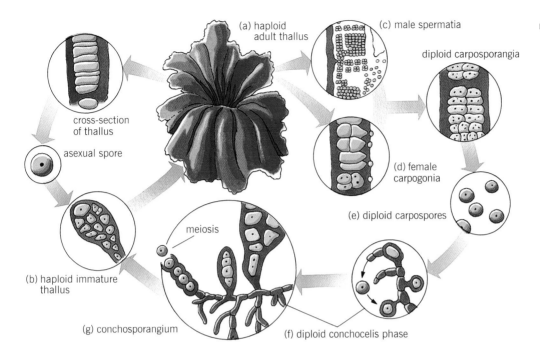

(a) haploid adult thallus

(c) male spermatia

cross-section of thallus

asexual spore

diploid carposporangia

(d) female carpogonia

(e) diploid carpospores

meiosis

(b) haploid immature thallus

(g) conchosporangium

(f) diploid conchocelis phase

LEFT **Life cycle of *Porphyra*.**

Carposporangia which in turn produce spores called carpospores (e). The carpospores are released and settle on to shells of molluscs or barnacles where they germinate into a filamentous structure that bores into the calcium carbonate (f). Before the work of Kathleen Drew of the University of Manchester, the connection between *Porphyra* and this filamentous stage was unknown, so much so that the filamentous stage was thought to be a different species, *Conchocelis rosea*. Dr. Drew published the link in 1949, and her finding led to the

FAR LEFT **Production of carpospores by *Porphyra*.**

LEFT **Adult *Porphyra* fronds.**

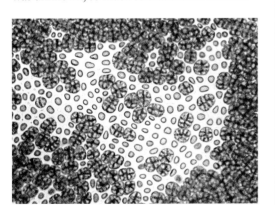

RIGHT **The conchocelis phase of the *Porphyra* life cycle grows on the surface of oyster shells.**

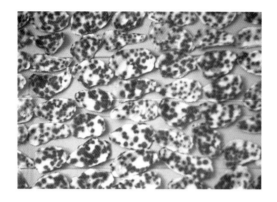

RIGHT **The conchocelis phase of the *Porphyra* life cycle grows on the surface of oyster shells.**

RIGHT **Many seaweeds such as this brown filamentous *Ectocarpus* release pheromones that cause the sperm to be attracted to the eggs.**

transformation of the *Porphyra* industry. With the complete life cycle known, the seaweed could be farmed more efficiently (see p. 86). The shell-boring phase is still called the *Conchocelis* phase (*concho* is Greek for shell) and this grows and reproduces asexually (by producing spores) to the extent that red smudges can be seen to cover the

shells. These sporophytes produce specialised structures, conchosporangia (g), in which meiosis takes place in some species to produce haploid spores (conchospores,) that settle and grow into the large haploid thalli (a), thereby completing the cycle.

Seaweed pheromones

As different as these three life cycles are, they all share the problems associated with reproducing in a turbulent sea. One of the most obvious hurdles is how to bring the sperm and eggs into contact. If they are simply released at random into the water there is a huge potential that fertilisation will be low. In many of the brown species the sperm are motile and are attracted to chemical signals, known as pheromones, released by the eggs. The highest concentrations of pheromones are

found close to the source, so by homing in on these chemicals, sperm can find their way to suitable eggs. Several pheromones have been characterised but they are not necessarily species specific. For example, the pheromone ectocarpene, the first to be identified in laboratory cultures of the brown *Ectocarpus siliculosus*, will attract sperm released from *Ectocarpus* as well as species of *Sphacelaria* and *Adenocystis*. Cross-fertilisation does not occur, however, due to features on the surface of the eggs that prevent fertilisation by the sperm of another species.

The monoecious brown *Sargassum muticum* adopts an alternative approach. It exudes copious amounts of mucus at the same time that sperm are released, while the oogonia are still connected to the thallus. This ensures that the sperm and eggs are not swept away by the sea, enhancing the chances of fertilisation, although the chances for self-fertilisation are also increased.

Mass spawning

Of course, synchronising the release of eggs and sperm helps to ensure that fertilisation can take place. This strategy is well documented for corals and gorgonians (sea fans and sea whips related to sea-anemones). Unlike these, spawning by most seaweeds does not appear to be controlled by lunar or tidal cycles. *Fucus* gametes are released from desiccated thalli on incoming tides. This can be extensive enough to turn the sea milky. Mass spawning of the green seaweeds *Caulerpa*, *Halimeda*, *Penicillus* and *Udotea* has also been shown to occur in predawn hours. Gametes are released in green clouds

or mucus streams lasting up to 15 minutes. The gametes remain motile for up to an hour and can swim up to 1 m (3 ft) within just a few minutes.

Fertilisation is not the only problem. The zygote must settle and attach itself swiftly before it is swept out to sea. The same is true for the settlement of asexually produced spores. As is often seen in the natural world, sheer numbers ensure the survival and maturation of at least a few specimens. When walking at low tide through a stand of sexually mature *Ascophyllum*, it is remarkable how fully laden the thalli are with receptacles. It has been estimated that a single *Nereocystis luetkeana* thallus can produce 3.7 billion spores over its single season lifetime. Each one would be capable of growing into a single gametophyte. However, the fact that we seldom see dramatic increases in standing crops of seaweeds from year to year is a clear indication of how enormously wasteful the reproductive process must be.

ABOVE **In an effort to ensure reproductive success seaweeds expend tremendous energy, as shown by the huge amount of receptacle production in this stand of *Ascophyllum nodosum*.**

How do seaweeds settle?

Due to the physics of turbulent water, any surface, such as a rock, the seabed, or even the surface of a seaweed, induces a boundary layer effect. This means that the movement of water very close to that surface is reduced to almost nothing at all. Safe within this boundary layer, spores, zygotes and fertilised eggs can easily attach and germinate. Attachment takes place with the help of chemical adhesives that harden with time and are enhanced by the development of small thread-like structures (rhizoids) that grow into crevices, cracks and around objects.

Many spores are non-motile and sinking to suitable substrata relies on physical forces. Motile cells are definitely more efficient at settling than their non-motile counterparts. Several seaweeds have enhanced the chances for settling. For example, the eggs of *Sargassum muticum* are fertilised when still attached to their conceptacles and develop into small multicellular organisms while still attached to the parent thallus. These then fall off, sinking considerably faster than a fertilised egg alone would have done. The Pacific sea palm, *Postelsia*, releases spores when the spray of a rising tide dampens the thallus. As the water droplets trickle down the grooved fronds, the spores are caught up into suspension and drip off the thallus on to the substratum below. The spores settle and attach in about 30 minutes, well before the tide covers them.

Even animals can help to enhance the settling rate. Grazing animals, such as copepods and amphipods, feed on tissue fragments, spores, fertilised eggs and zygotes in the open water. Seaweed spores and fertilised eggs can often survive the passage through the animals' guts and are passed out within faecal pellets that are comparatively heavy and sink rapidly. The faecal pellets help protect the new algae from desiccation in the intertidal zone of the beach and, because they are sticky, aid attachment. Just as seedlings grow well on a compost heap, the faecal pellets also supply any developing seaweed with a rich source of nutrients.

BELOW *Laminaria* sporelings produce fine rhizoids that initially attach the fronds to rocks and other solid objects.

BELOW RIGHT Spores of the brown filamentous epiphyte *Elachista fucicola* settling on the surface of the brown *Fucus vesiculosus*.

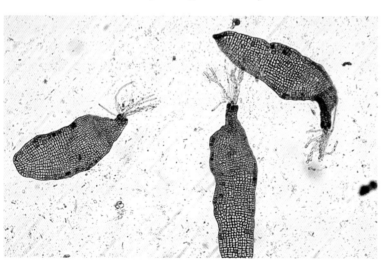

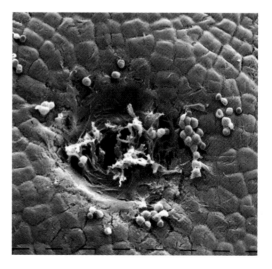

RIGHT **The distribution of young *Fucus* plants on this boulder exemplifies the patchy distribution of individuals following settlement.**

BELOW **Some seaweeds grow in unattached forms such as *Jania rubens* that forms balls which can accumulate in large numbers as here in shallow waters near Abu Dhabi.**

Seaweed balls

Of course, not all seaweeds need to attach themselves and the free-floating forms can be quite different in structure and growth patterns. As the thalli are tumbled in the water, they grow in all directions leading to the formation of compact balls. Although this has been shown for several different seaweeds, including the brown *Ascophyllum*, red *Chondrus* and the coralline *Jania*, the most well known case is actually a freshwater species called *Cladophora aegagropila*. This green seaweed forms tight balls about 10 cm (4 in) in diameter that even have a place in local folklore and are a focus of a summer festival of the Ainic people in the Hokkaido district of Japan. The balls are call 'marimba' and the mythology involves a young girl and man who drowned in the lake, their hearts turning into the algal balls. Curiously, when a species becomes free-floating, reproductive structures seldom form and the seaweed can only reproduce by fragmentation.

Free-living corallines

Species of unattached red coralline seaweed do not float due to their weight. Several species, such as *Phymatolithon calcareum* and

Lithothamnion corallioides, form extensive beds of fragmented calcareous irregular nodules called maerl on the sea floor at depths of 3–25 m (11–82 ft). Other non-articulated, free-living coralline algae, including *Lithothamnion* species, are found at depths of 50–200 m (164–656 ft). These nodular structures, often also called rhodoliths, are extremely slow growing and deep-water forms take up to 800 years to reach a diameter of 30 cm (12 in).

How long do seaweeds live?

Many seaweeds are perennial and can live for tens of years. *Ascophyllum*, for example, can live in excess of 30 years and *Macrocystis* between five and ten years. Some perennials, such as *Alaria* and *Catenella*, don't survive throughout the year as the complete thallus; only the basal structure is perennial while the upright fronds are produced annually or seasonally. A curious variation is that of the brown *Himanthalia elongata* which lives for two or three years as a short button-like disc, just a couple of centimetres across, on the shore. It reproduces only once, when it produces whip-like receptacles up to 2 m (6.5 ft) long.

Other seaweeds are much shorter-lived annuals. Some, such as the sheet-like *Porphyra*, *Ulva* and *Enteromorpha*, live just a matter of weeks. Somewhat surprisingly some of the large brown species, such as *Saccorhiza polyschides*, are annuals that grow very fast. Amazingly *Nereocystis* is also an annual that manages to grow taller than many mature trees within a single season.

Triggering reproduction

The timing of reproduction and growth of seaweeds is closely linked to environmental triggers such as temperature and day length. The alternate generations of a species often thrive under different environmental conditions: when conditions are right for one form, vegetative growth and asexual reproduction will take place; when conditions change they trigger the seaweed to move into its sexual phase. There is an increasing body of research to show that some seaweeds

LEFT **Maerl beds form extensive reefs that are important refuges for a diverse invertebrate fauna.**

BELOW **The timing of reproduction in seaweeds is controlled by a combination of several factors such as day length and temperature.**

follow yearly (circannual) rhythms in their patterns of growth and reproduction. These rhythms are governed by seasonally changing environmental cues and species that respond in this fashion are called 'season anticipators'. These contrast to 'season responders' that grow and reproduce only when environmental conditions are suitable.

Usually several triggers work in unison. Day length alone, for example, is a poor environmental trigger since short day lengths occur in spring, autumn and winter. Growth and reproduction responses are sometimes only induced when critical day lengths take place within a certain temperature range. For example, in *Porphyra tenera* the formation of conchospores (see p. 26) is promoted by eight hours' illumination per day but inhibited by day lengths over 14 hours. There is also an optimum temperature of 21°C (70°F) for the formation of the spores. Therefore, on the coast of Japan where the species grows, the greatest quantities of conchospores are released towards the end of September when day lengths are below 10 hours and the water temperature is below 22°C (72°F).

The rigours of seaweed life

The region between the open ocean and the land is a demanding place in which to live. It is a place of extremes dominated by tides and waves. Seaweeds have to be able to cope with the rise and fall of the tides. Depending on their position on the shore, they may have to spend hours out of the water, or they may be trapped in rockpools, where the environment can change suddenly with a downpour of rain. Yet seaweeds have adapted so that they can thrive in places where it would seem impossible for any immobile organism to survive.

Tidal patterns

The tidal patterns of a particular stretch of coast determine which species of seaweed can survive there. There are three main classes of tidal cycles: diurnal, semidiurnal and mixed. Diurnal tides, with one high and one low water per day, are characteristic for parts of the Gulf of Mexico. Semidiurnal tides, with two high and two low tides in a 24 hour period, are characteristic of Atlantic coastlines. Mixed tides found on Pacific and Indian Ocean coasts, as well as small basins like the Gulf of St Lawrence, also occur twice

LEFT **The periodic rise and fall of the tide is fundamental for determining where seaweeds grow on the shore.**

ABOVE **Some seaweeds such as *Laminaria* are only exposed to the air at low spring tides; normally they remain totally submerged.**

will be 20% greater than the monthly average. Likewise, twice a year, as the earth reaches its closest position to the sun in June and December, the greatest tidal amplitudes are normally experienced.

Tidal ranges vary greatly with location. In some regions tidal ranges seldom exceed 1 m (3 ft), whereas in the Bay of Fundy in Canada the spring tides range up to 17 m (56 ft) and the Severn Estuary in England experiences 13 m (43 ft) spring tide ranges. However, unpredictable storm-induced events have significant influences on any tidal ranges that are normal for a particular site.

Seaweeds and tides

Any seaweed that grows on the shore between the lowest and the highest tide levels will be exposed to the air at some point, maybe once or twice a day. However, for those that grow at lower levels exposure to air is a rare event. For most of the time these seaweeds grow in seawater of reasonably constant salt composition, although there may be seasonal changes in nutrients and temperature. Higher up on the shore, however, conditions are quite the opposite: seaweeds may face long periods out of water and must be able to cope with the full force of the summer sun as well as harsh winter frosts. In rock pools the saltiness (or salinity) of the water may increase as water evaporates, or dramatically fall if it is raining or snowing. Of course, the timing of the low tide has important consequences: in hot seasons low tides during the day result in far more extreme conditions than low tides at night; whereas in cold weather the reverse is probably true.

a day but the highs and lows are of unequal amplitudes. The Baltic Sea is an example of an atidal sea with a tidal range of less than one metre.

What causes the tides?

Tides are produced by the gravitational effects of the moon and to a lesser extent the sun. Monthly changes in the alignment of the moon, sun and earth result in two major tidal classes: spring tides and neap tides. Spring tides take place when the moon and sun are in line with the earth, which occurs twice a month at the new and full moon. Neap tides occur when the moon, earth and sun are at right angles and have, on average, 20% lower amplitudes than spring tides. The elliptical orbits of the moon and earth also result in twice monthly events: when the moon is at its farthest point of orbit, the tides will be 20% less than average and when it is at its closest point of orbit, again twice a month, the tides

Zones on the shore

Phycologists studying the distribution of seaweed species on the shore have found that certain species are specific to particular regions or zones. The zones are delineated by critical tide levels that govern how long the region is submerged or exposed. Whether a species can grow in a particular zone depends on its tolerance to desiccation, temperature stress, salt stress and wave action. As well as tidal patterns, the positioning of such zones on any shore is greatly influenced by the degree of wave action routinely occurring there. On more exposed shores the wave action is greater and so spray will splash higher up the beach than on a sheltered shore with fewer waves.

The shore that lies between the extreme high water of spring tides and the extreme low water of spring tides is called the intertidal or littoral zone. The supralittoral is the region above the littoral zone that receives spray and the sublittoral or subtidal zone is the permanently submerged region that lies beyond the extreme low watermark. The depth to which the subtidal zone extends is largely governed by the limits of light penetration.

The intertidal zone

Worldwide studies have identified three regions within the intertidal zone. There is an uppermost strip of desiccation-tolerant lichens (normally the black *Verrucaria*) or cyanobacteria that often have obvious populations of littorinid snails. The next zone down the shore comprises a wide diversity of seaweeds together with barnacles and limpets. The lowermost part of the intertidal zone harbours a diverse community of seaweed species including red encrusting and coralline algae.

The subtidal zone

The kelps of temperate and some tropical shores grow within this zone. The lowermost region of this zone is never exposed to the air and is dominated by a variety of large canopy-forming seaweeds with diverse understorey communities of predominantly red and brown species. Still further down, where light hardly penetrates, is home to encrusting red algae.

The problems of desiccation

Seaweeds living in the intertidal zone may face long periods out of water, but there are various strategies for keeping water loss to a minimum. Even when seaweeds do dry out they have a remarkable ability to recover once the tide returns.

BELOW **On exposed shores wave action will exert a greater influence on the distribution of seaweeds on the shore than on sheltered shores.**

LEFT Intertidal zones are characterised by having distinct patterns where species of seaweeds and animals inhabit distinct bands on the shore.

LEFT On tropical shores dense growths of cyanobacteria are found high up on the shore where lichens grow in colder regions.

Safety in numbers

On many shores at low tide you will see swathes of seaweed exposed to the air. However, lift the thalli up and there is likely to be many layers of the same species beneath. Beneath these there will be other species forming an understorey layer and beneath them there may be further encrusting or turf-like species growing on the substratum. The consequence of this sort of layered structure is that it is only those fronds lying on the surface that will be exposed to the harsh conditions. The underlying fronds and species will be buffered, so that on a warm summer's day although the surface layer is being roasted, the seaweeds beneath it are in a warm, humid environment, probably similar to the conditions in a rainforest. The surface layer acts as an effective barrier to the evaporation of water, especially once it has dried out. Likewise in freezing temperatures, the surface fronds may be frozen stiff, but the understorey thalli will remain cold, but unfrozen.

Even within a single species the same principle applies. Tufts of *Cladophora* or *Ectocarpus* comprising of branched filaments are often found hanging from rock faces. When the tide goes out the tufts actually retain a lot of water between the filaments so that even after they have been exposed to the air for a considerable time the interior of the tuft remains moist. Close turfs of algae that are commonly seen on tropical shores also retain water in this way.

LEFT **The fronds lying on top of other seaweeds form an effective barrier preventing water loss and extreme temperature fluctuations in the underlying layers.**

Preventing water loss

Desiccation can be extreme such that intertidal seaweeds can lose up to 90% of their tissue water when exposed to the air, especially if there is a breeze and humidity is low. The speed with which this happens depends largely on the surface area to volume ratio of a thallus. The larger this ratio, the faster the water loss; in general, the smaller seaweeds, which have higher ratios, are more prone to such losses. The more branched a thallus is, the more surface area there is from which water can evaporate. Therefore the highly branched *Ascophyllum* loses water more readily than the less branched *Laminaria* thallus. Broad, flattened sheets of *Ulva* and *Porphyra* lose water far more quickly than a tubular thallus of *Enteromorpha* of the same mass.

Fucus* and *Laminaria* species may also withstand a certain amount of desiccation due to the presence of polysaccharides within their inner cortex and medulla tissues that bind with water. Other seaweeds exude

mucus that helps to slow down the rate of evaporation. If you walk across a patch of *Halosaccion glandiforme*, a widespread red seaweed on northern Pacific rocky shores, you will be hit by squirts of water since this species actually avoids desiccation by storing water inside its robust sac-like thallus.

Just add water

The most obvious factor that determines whether or not a species can survive long periods out of the water is whether it is able to photosynthesise in air. In the water seaweeds obtain the carbon needed for photosynthesis from dissolved carbon dioxide (CO_2) or bicarbonate (HCO_3). When they are exposed to air, photosynthesis can only take place with the uptake of carbon dioxide from the air. As long as the seaweeds do not dry out, many species photosynthesise in air at rates similar to those measured when they are fully submerged. However, as they begin to dry out their ability to photosynthesise diminishes. Photosynthesis in *Laminaria*, even when mildly desiccated, is greatly reduced whereas *Ulva* continues photosynthesis down to 35% water loss. *Fucus vesiculosus*, *F. serratus* and *F. spiralis* all photosynthesise when desiccated down to tissue water contents of less than 30%.

When covered by a rising tide, many of the species mentioned recover their full photosynthetic rates within hours. *Pelvetia canaliculata* is found very high on the shores of Europe and is prone to drying out for long periods of time. Researchers found that within less than a day of being back in seawater, a specimen that had been desiccated

ABOVE **The brown** *Pelvetia canaliculata* **grows high on the shore where it is particularly tolerant of long periods of desiccation.**

for six days was able to resume full rates of photosynthesis. Even seaweeds left out of water for 28 days were photosynthesising at 20% original levels after four days back in water. In fact, *P. canaliculata*, requires periods of exposure to the air. If it is submerged for more than six hours out of 12 it actually starts to decay. This is a rare example of a seaweed species in which periods out of water are absolutely essential.

Dealing with salt stress

In most oceans around the world there are about 35 grams of salt in every litre of water. This is referred to as the water having a salinity of 35. There are notable exceptions to

this, such as the Baltic Sea, which has much reduced salinities of between three and 10 due to its large influx of freshwater, while the Arabian Gulf, Mediterranean and Red Sea have salinities up to 45 or more due to high rates of evaporation and their semi-enclosed nature. The predominant salinity of any shore depends on whether there is river discharge nearby and how rapidly this discharge is mixed with ocean water, although in general coastal waters have reduced salinities compared to the open sea.

When seaweeds are exposed to rain at low tide, they experience a dramatic reduction in salt concentration. In contrast, when water evaporates from the surfaces of seaweeds, salts in the remaining water become greatly concentrated. Both lowered and elevated changes cause a severe strain on the cells of the seaweeds, requiring swift metabolic action to prevent permanent damage. Of course, sometimes the evaporation of water on the shore is so great that all that remain are salt deposits; seaweeds are unlikely to survive in such conditions.

Salt and osmosis

In normal seawater the cells of a seaweed are in balance with the water outside of the cells. When salinity in the external water goes down, it creates an osmotic imbalance that results in water being taken up by the cells (the amount of water taken up is in proportion to the reduction in external salt concentration). The cells combat this uptake of water by lowering the cellular concentrations of ions such as potassium, sodium and chloride, as well as sugars and organic compounds to restore the osmotic equilibrium to one close to that before the salinity had been reduced. When the salinity of the external water increases the reverse process takes place. The higher concentration of external salts causes water to be lost from the cells, again in direct proportion to the increase in salinity. The cells have to restore the osmotic balance by taking up ions, or producing greater cellular concentrations of sugars and organic compounds. These sugars and organic compounds are many and varied, including mannitol, sucrose, floridoside, isofloridoside, proline, glycine betaine and a curious sounding compound called dimethylsulfonioproprionate or DMSP for short.

DMSP is a compound gaining much recognition these days because it is a precursor for the volatile gas dimethylsufide, DMS. The latter is released when DMSP in the cells of red and green seaweeds (browns don't produce it) is broken down. DMS in the atmosphere is oxidised to a variety of chemicals that act as cloud condensation nuclei. High accumulations of DMS from seaweeds can play an important role in cloud formation and therefore in localised and large-scale climate regulation. It is DMS that, together with ozone, creates that wonderful 'seaside aroma'.

Salinity tolerance varies

Seaweeds have differing capabilities for making these osmotic adjustments and hence

LEFT **In some regions, such as this Moroccan estuary, the evaporation of water is so extreme that no seaweeds can survive.**

LEFT **Shallow rock pools high on the shore can quickly turn to very concentrated brine pools where only the hardiest of seaweeds, such as the green *Enteromorpha* seen here, can survive.**

varying tolerances to salinity changes. In general, seaweeds growing higher on the shore tend to have greater tolerance than species found lower on the shore. Many intertidal seaweeds may withstand salinities ranging from 0–100, whereas many subtidal species are killed by only slight changes in salinity, or at best have limited tolerance of salinities from 20–50.

However, just because a particular species can withstand a particular salinity does not mean that it is able to photosynthesise or grow at normal rates: metabolic rates are severely reduced for most species in extremely low and high salinities. The biochemical pathways that bring about the osmotic changes in ions and/or organic compounds require energy from photosynthesis. Many of the responses are very reduced, if they take place at all, in darkness and again clearly the consequences for the seaweeds experiencing salinity changes during the night are very different to the responses that are possible during the day.

The rate of salinity change has also been shown experimentally to affect how well seaweeds such as the green *Cladophora* and *Ulva* are able to adapt. When changes are abrupt, the cellular osmotic adjustment does not take place as effectively as when the changes are gradual. Translating this to how the seaweeds may respond on the shore, when a light drizzle gradually lowers the external salinity, osmotic adjustment may be more complete than that taking place when a torrential downpour drenches the thalli.

Even species where the gametophyte and sporophyte stages are very similar, such as the filamentous brown *Ectocarpus*, they may have very different salinity tolerance. In the case of *Ectocarpus* the haploid gametophytes have a very narrow tolerance confined to salinities close to 34, whereas the diploid sporophytes are tolerant of a broad range of external salinities ranging from 16–68.

Effect of temperature on saline stress

As well as light levels, temperature can have a significant influence on how well seaweeds survive saline stress. Species normally grow within a certain temperature range and at temperatures at either extreme of this range the ability to withstand salinity stress is generally reduced. To confound the complexity even further juvenile and adult thalli of a species may have different tolerances to saline stresses. Young red *Delesseria* have a broader tolerance of salinity change over a wider range of temperatures than the adult forms of the same species.

Temperature stress

The geographical distribution of seaweed species is largely dictated by temperature tolerance. Consequently there are species of seaweed that are restricted to tropical waters with narrow temperature tolerances close to 30°C (86°F), while species in the Antarctic rarely survive temperatures above 13°C (56°F).

LEFT **Frozen fronds of** *Fucus vesiculosus* **less than an hour after being uncovered by a falling tide.**

Different life cycle stages may also be sensitive to different temperature ranges, i.e. the temperature limits for survival of a gametophyte may be different to the preferred temperature limits of a sporophyte of the same species. Temperature tolerance, and therefore geographical distribution, is also modified by the ability of a species to withstand other stresses, such as salinity and desiccation, especially at extreme high and low temperatures.

Seasonal temperature changes

Seasonal changes in water temperatures are largely gradual and species that live permanently submerged are not often exposed to sudden shifts in temperature. Metabolic rates change with these seasonal variations due to the effects of temperature on chemical reaction rates, enzyme metabolism and biophysical processes. At the lower end of the temperature range for a particular species, growth rates will be slower than at higher temperatures, and often there will be temperatures at which growth is optimal although these may vary with the different life cycle stages. When temperature conditions become unfavourable for one life cycle stage, the initiation of the next generation may be induced, as in the example given below for *Porphyra*.

A hot summer

In the intertidal zone seaweeds exposed during periods of low tide may be subjected to temperature extremes way outside the

LEFT **On a hot afternoon these exposed *Porphyra* fronds rapidly reach temperatures far greater than the water in which they were bathed just a few hours earlier.**

normal limits set by seawater temperatures. On a hot, cloudless summer's afternoon high on the shore it is not difficult to imagine a dark, radiation absorbing seaweed thallus quickly reaching high temperatures. For example, temperature measurements have been made of a *Porphyra* thallus on a calm sunny day, where the water temperature was 13°C (56°F). When the water fell from the thallus at 9 am its temperature quickly rose to 17°C (62°F), the temperature of the air. Over the following 5½ hour low tide period the temperature of the thallus rose to 34°C (93°F). Subsequently when the incoming tide covered the seaweed, the temperature of the thallus plummeted to 13°C (56°F) immediately. These are harsh temperature shifts for any organism to survive.

A freezing winter

As with salinity and desiccation tolerances, it is evident that intertidal species are in general more tolerant of temperature fluctuations than subtidal species. This also applies to shores where freezing temperatures occur. Most seaweeds would be killed if frozen. However, high concentrations of tissue salts and organic solutes in the seaweed's cells lower the freezing point, lending a degree of protection. *Fucus vesiculosus* can withstand temperatures of –40°C (–40°F) for several months and *Ascophyllum* can survive in temperatures down to –20°C (–4°F) with no indication of injury. Some *Porphyra* species can withstand freezing down to –70°C (–94°F) for up to 24 hours.

Differences in freezing tolerance play a role in determining the distribution of

seaweeds on the shore. In New England and eastern Canada *Mastocarpus stellatus* is routinely found higher on the shore than a similar species, *Chondrus crispus*, and therefore exposed to freezing temperatures for longer. Experiments with the two species have shown that whereas *M. stellatus* can withstand freezing at –20°C (–4°F), *C. crispus* thalli are badly damaged at such low temperatures.

ABOVE **A group of *Fucus* thawing in the early morning sun, after being frozen during a low tide the night before.**

LEFT **In Canada and New England the red** *Chondrus crispus* **shown here is more susceptible to freezing temperatures than a very similar seaweed called** *Mastocarpus stellatus*.

As with salinity stress, the rapidity of the temperature shock may be critical in determining whether a thallus survives. The similarity between high salinity stress and freezing stress doesn't end there, since several of the organic compounds that help to regulate saline balance (such as DMSP and proline) are thought to act as antifreezes and osmotic buffers within the cells of some seaweed species.

Water movement

Another challenging aspect of seaweed life is the relentless action of the waves. The water crashing on to a shore is problem enough, but when the waves pick up sand they become like industrial sandblasting machines. In particularly violent waters, such as those following storms, pebbles and stones replace the sand, making the waters even more efficient scouring agents. In the parts of the world where coastal waters freeze over, lumps of ice suspended in the water can be just as devastating. Even the most stubborn growths can be stripped from a rock face.

The turbulent motion of seaweeds that is induced by currents and waves is not all bad,

BELOW **Incoming tide straightening** *Laminaria* **fronds.**

RIGHT **The ruffles and uneven surface of the blade of *Laminaria saccharina* induces localised water turbulence thereby increasing the exchange of gases and nutrients.**

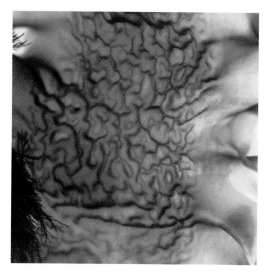

however. The currents and waves of coastal regions mean that seaweeds are constantly supplied with nutrient-rich water (water can quickly become depleted where movement is low and where large stands of seaweed are present). Constant water motion also distributes light in the water more evenly. Ripples on the surface of the water can focus light to spots of harmful intensities, but because the ripples are constantly moving around, these intense bursts of light do not damage the organisms living in the water below.

In regions where wave action is strong, grazing by herbivores is significantly reduced. Turbulent water also keeps the seaweed in constant motion, allowing the diffusion of nutrients and gases to and from the fronds. *Nereocystis* has smooth blades in rapidly moving water, but in more sheltered waters its fronds are ruffled. The ruffled blades serve to increase the turbulence as water passes over them thereby increasing nutrient supply and gas exchange to the fronds. In faster moving waters the ruffled blades would increase the risk of tearing because of their increased drag; instead the seaweed's smooth blades tend to form streamlined bundles that are not so easily damaged.

It pays to be flexible

A key to the success of the seaweeds is their flexible nature, especially those with long stipes. If the seaweeds were like trees or bushes, with rigid woody stems, they simply would not survive in turbulent waters. The energy of the currents and waves is far greater than severe winds and rigid structures would be broken and displaced too easily. Instead, the flexible stipes allow the thalli to bend in the direction of the water movement. Water passing over an assemblage of seaweeds such as a dense *Laminaria* forest, tends to flow over the group rather than around the individual thalli. In severe conditions this may prevent considerable damage.

Seaweeds adapt their shape to reduce drag depending on their location. The same species growing on exposed or sheltered stretches of coast can look quite different. In general, wave-exposed forms are smaller and may have narrower blades. *Fucus vesiculosus* thalli from exposed sites are shorter, thinner and have fewer air bladders than those from sheltered shores.

Growth strategy

Despite having smooth blades and a flexible stipe, seaweeds suffer considerable damage from surf and wave action. The fronds of species such as *Durvillaea* and *Laminaria* are frequently broken off at the tips or lacerated

RIGHT **The long flexible, hollow stipe and lacerated blades of** *Ecklonia maxima* **make it ideally suited to growth in turbulent South African coastal waters.**

RIGHT **The sea palm** *Postelsia* **is a tenacious species that is able to grow high on shores with severe wave action.**

into strips. Fortunately many of the kelp species have a growth pattern that allows them to overcome this problem. New growth occurs in the meristematic region where the frond joins the stipe. Therefore loss of tissue from the tip of the thallus can be replaced by new growth from lower down.

One of the most tenacious seaweeds to survive high-energy wave environments is the annual sea palm, *Postelsia*. It grows along the eastern pacific coast from British Columbia to California on rock faces in high to mid-intertidal zones that are buffeted by very heavy wave action. The secret of its success seems to be that it has exceptionally strong holdfasts combined with highly flexible stipes. The possession of effective holdfasts and strong adhesives is key to survival in these high-energy environments. However, this is only true if the holdfasts are attached to solid rock surfaces. Often drift seaweeds are washed up, where it is soon appreciated that the attachment to stones, mussels or even fair sized boulders is not enough to guarantee a permanent foothold.

The problem of sand

Sand scouring, by water laden with sand certainly cleans rock surfaces. Rocks in these regions tend to be dominated by quick growing opportunistic species that reproduce swiftly. However, there are some tough seaweeds that are highly tolerant of sand scour and can even survive being buried in 1–2 m (3–6.5 ft) of sand for several months at a time. The red *Ahnfeltiopsis linearis* and *Ahnfeltiopsis concinna* and the brown *Laminaria sinclairii*, amongst others, grow on rocks low in the intertidal zone of sandy areas. They time their growth and reproduction for periods when they are uncovered, before seasonal depositions of sand cover the thalli.

Population variation

Continued exposure to different environmental conditions may lead to distinct populations of the same species having very different growth requirements. Examples already discussed are the different shapes of *Fucus* thalli found on exposed and sheltered shores. However, such differentiation can also be seen in the physiological responses to factors such as salinity, temperature and light. Populations with different physiological capabilities, even on a single shore can be established. For example *Cladophora rupestris* populations from high on the shore (which are prone to low salinity deluges by rainwater) are far more tolerant of low salinity than the same species found growing lower on the shore.

BELOW **Dense growth of** ***Fucus*** **in a Moroccan estuary. Because of the high rates of evaporation, the salinity of the water is very high and to survive, this population is physiologically adapted to much higher salinities than populations in less saline estuaries.**

Estuarine populations

On a larger scale such population differences have been shown for estuarine populations of the brown filamentous *Pilaiella littoralis*. Samples collected from virtually freshwater stretches of the estuary had very different salt tolerances to those collected from close to the mouth of the estuary where seawater was the main influence. These differences are genetically controlled, since the trends are maintained in the progeny of the populations even when grown under controlled conditions in the laboratory for many generations. Samples of another filamentous brown species, *Ectocarpus siliculosus*, collected from different locations around the world still demonstrate considerably different salt tolerances even after being grown in the laboratory (all at the same salinity) for over 30 years. The same *Ectocarpus* samples also show differences in their response to temperature stress. The populations have maintained their distinct preferences for certain temperature ranges despite the fact that they have all been grown for three decades at one temperature.

It is generally populations of marine seaweeds with low salinity tolerances that extend up into the freshwater influenced parts of estuaries. Very few freshwater species extend in the other direction, the only commonly described species being *Cladophora glomerata*.

BELOW *Enteromorpha* **growing on the banks of an estuary, with freshwater rivulets flowing over the mats in the foreground. In an estuary such as this, distinct populations of a particular species will exist with varying salt tolerances.**

Until recently it was thought that separate freshwater and marine species of the red filamentous genus *Bangia* were confined to either full seawater or freshwater, with no overlap along estuaries. However, freshwater *Bangia atropurpurea* can be acclimated to grow in full seawater after several generations, and marine *B. atropurpurea* can be acclimated to freshwater.

Estuaries and salt marshes are often very muddy places with large stretches of mud flat exposed at low tide. Low salinity tolerant populations of species such as *Enteromorpha* can form dense growths that help to stabilise the mud surfaces. Sometimes these dense growths are buried by fresh mud, but survive to form a dormant population that is ready to flourish when the surface mud is disturbed again.

Light

In the clearest seawater (found in polar or tropical regions) light may penetrate 200 m (656 ft). But for most waters light penetrates tens of metres rather than hundreds, and in some instances far less than that. Divers in coastal waters often testify that they can hardly see their hands in front of their masks, especially following a storm. In fact over 90% of the ocean floor is in permanent darkness and no seaweed growth is possible.

Of course, seasonal changes in day length and the angle of the sun dramatically alter the amount of light seaweeds receive. But there are other factors that play a role in reducing the amount of light reaching depths. Between 4 to 30% of the incident light is reflected from the sea's surface. Small waves can lower

ABOVE **The depth to which light penetrates is key to the growth of seaweeds in water.**

LEFT **Many water properties affect the transmission of light, and even small ripples on the surface can alter the light reaching the seaweed below, focusing the light into bright points of high intensity.**

the amount of light that is lost in this way, but when small bubbles form in more turbulent waters as much as 50% of the incident light can be reflected. Particles in the water, such as suspended sediment, plankton, and even dissolved organic matter (DOM) all combine to further prevent light filtering through. Seawater effectively absorbs the red part of the light spectrum and far red light penetrates only the very top metres. The blue part of the spectrum is also absorbed by the particles and DOM. Therefore, as light penetrates deeper, not only does the amount of light diminish, but the spectrum also narrows with depth.

Seaweed pigments

Seaweeds have different pigment compositions that give them their characteristic colours. The pigments also enable seaweeds to optimise their ability to capture light at different water depths. The pigments absorb light energy between the wavelengths of 400–700 nm, the section of the light spectrum used for photosynthesis. All seaweeds contain the green chlorophyll a that absorbs light at wavelengths between 400–500 nm (green) and 650–700 nm (red). Other types of chlorophylls and pigments absorb light in other parts of the spectrum: carotenoids such as ß-carotene and fucoxanthin (found in brown seaweeds) and chlorophyll b (found in the green seaweeds) absorb in the green part of the light spectrum (400–520 nm); phycoerythrin (found in red seaweeds) absorbs in a different part of the green region (490–570 nm); phycocyanins and allophycocyanins (found in red seaweeds) absorb light in the green-yellow (550–630 nm) and orange red (650–670 nm) parts of the spectrum, respectively.

Growing in the depths

It is generally accepted that seaweeds need minimum irradiances of 0.05–0.1% of the surface light to be able to grow. However, red encrusting seaweeds have been found growing on a seamount in the Bahamas 268 m (879 ft) deep, where the light has been estimated to be about 0.0005% of surface light levels. This is about the same intensity as the light hitting the surface of the sea on a moonlit night. Seaweeds that grow in low light conditions are generally shade adapted, which means that their photosynthetic metabolism and light-absorbing pigments are highly efficient. For example, populations of the red *Chondrus* growing at depth tend to have increased concentrations of the red pigment phycoerythrin compared to populations growing in better lit positions on the shore. It used to be said that it was the red species that were the only ones adapted to growth in the deep waters, although green and brown species are being discovered growing at water depths of 50–100 m (164–328 ft).

Shape and photosynthesis

Rates of photosynthesis are also greatly influenced by the shape of the seaweed, the surface area to volume ratio, and the amount of non-photosynthetic tissue. Sheet and tubular seaweeds, such as *Ulva, Porphyra, Enteromorpha* and *Dictyota*, have the highest rates of photosynthesis. Finely branched filamentous algae, including *Cladophora, Ectocarpus, Ceramium* and *Chaetomorpha*, are the next most productive. Their tendency to clump together, so that the interiors of the clumps are shaded from light, is what makes them less productive than the sheet-like seaweeds. Yet they are far more productive than the coarsely branched species such as *Gigartina, Codium, Chondrus* and *Mastocarpus*. Encrusting algae and calcified species have the lowest rates of photosynthesis, especially those with thick calcium carbonate depositions that are very

RIGHT **High light and ultraviolet radiation can be damaging to seaweeds exposed at low tide or which float in surface waters.**

effective barriers to light. Although having higher productivity rates than the coralline species, thick-bladed seaweeds, such as *Laminaria, Fucus* and *Sargassum*, also have comparatively low productivity rates because they generally have a large amount of non-photosynthesising tissue. In such thalli, there is a significant variation between tissues of different ages: in a blade of *Macrocystis*, for example, the thin tips have a greater proportion of photosynthetic tissue than the thicker base of the blade.

Ultra violet protection

Many species of seaweed are damaged or inhibited by high light levels, particularly high levels of ultra violet radiation. Ultra violet radiation can damage cellular proteins, RNA transcription, DNA replication and photosynthetic mechanisms. To combat the effects of ultra violet radiation, seaweeds produce protective screening agents, including amino acids and carotenoids. Ultra violet damage is most prevalent in seaweeds that are exposed at low water, as well as those growing in shallow waters. In coastal seawaters ultra violet penetrates only a few metres at most.

Rock pools

Rock pools are a conspicuous feature of many rocky shores during low tide. They may be no more than puddles that rapidly evaporate or large ponds with surface areas of several square metres and depths of over a metre. Conditions prevailing in any pool depend on the surface area to volume ratio, the height on the shore and how long the pool is cut off from the sea. Deep pools can have diverse seaweed communities that resemble those of the lower shore, even showing distinctive zonation patterns. Shallower pools, especially those higher up the shore tend to be populated by less diverse communities of species.

The problems of rock-pool life

In general the pools can be considered to be refuges from the desiccation, salinity and temperature stresses already discussed. However, they are not stress-free habitats. On a hot day a pool warms up and evaporation of water increases its salinity, while on a rainy day low saline freshwater may collect on the surface of the pool (since freshwater floats on top of seawater). All these events can set up gradients of temperature and salinity with depth throughout the pool. Pools high on the shore can freeze over in extremely low temperatures, but since only pure water can form ice crystals the layer of ice on the surface of a pool results in significantly elevated salinities in the water below.

Pools, especially those high on the shore, tend to collect seaweed debris. If this is not washed away it decays and uses up all the oxygen in the water, leading to the production of foul-smelling hydrogen sulphide gas.

Just as in the open water light penetration is fundamental to where a species can grow, the same is true within rock pools. These tidal pools also help to protect some populations from ultra violet radiation damage. In the open water, or even in the open air, carbon dioxide and oxygen produced and taken up during photosynthesis are quickly replaced due to water movement and exchange.

However, in a pool where there is only a finite quantity of these gases, a stress is introduced that is not experienced by seaweeds growing on the open shore. During the day when photosynthetic rates are high, carbon dioxide

ABOVE **Rock or tide pools can be important refuges from the physical stresses of a rocky shore.**

RIGHT **Dense growths of *Enteromorpha* on the surface of a rock pool. The thalli have been lifted to the surface of the water by the copious amounts of oxygen produced during photosynthesis.**

is taken up and oxygen produced. On a warm summer's day you may see tiny gas bubbles rising from the surfaces of the seaweeds through the water of a pool. Oxygen is produced and carbon dioxide taken up causing a rise in pH. During the night the respiration of the seaweeds takes up oxygen and releases carbon dioxide, resulting in a lowered pH. If there is a lot of seaweed biomass in a small volume of water, such pH changes can be highly significant, and an additional stress for both the seaweeds and animals sheltering in the pools.

It is not just the seaweeds that inhabit such pools. Sit quietly beside most pools for just a few minutes, and numerous snails, urchins, crustaceans and even fish show themselves from behind the cover of the algal fringes. One of the downsides for seaweeds growing in such a pool is that they may be forced into a closer association with potential grazers than would be the case on the open shore.

Seaweed as part of an ecosystem

The total weight, or biomass, of seaweed on a shore can be impressive, especially when the subtidal kelp forests are considered. The productivity of algae and terrestrial plants can be expressed as the amount of carbon taken up through photosynthesis to produce new biomass per unit of area (typically per square metre) per unit of time. When compared in these terms, the productivity of seaweed communities is equal to, or in many cases greater than, that of terrestrial plant-based systems. For instance *Laminaria*-dominated communities have annual productivity rates of approximately 2 kg carbon per m² (0.4 lb per ft²) and *Postelsia* communities have been estimated to produce up to a massive 14 kg carbon per m² (2.8 lb per ft²). These estimates contrast to the yearly productivity rates of intensive alfalfa crops (1–2 kg carbon per m² or 0.2–0.4 lb per ft²) and temperate tree plantations or grasslands (generally less than 1 kg carbon per m² or 0.2 lb per ft²). Even the productivity of microscopic unicellular algae in the sea, that can grow so fast that they turn coastal waters from clear to a murky brown overnight, only reach annual production values of around 1 kg carbon per m² (0.2 lb per ft²).

Seaweed as a food source

Seaweeds are part of a complex ecosystem involving many different types of animals. Of course, many of the finer seaweeds are valuable food sources, yet the larger species are not grazed on to a great extent. In the case of *Laminaria*, for example, only about 1% of the thallus is directly eaten by herbivores such as gastropods, sea urchins and fish. Bits and pieces constantly break off from the main thallus and provide food for debris feeders such as crabs, sea cucumbers and amphipods. The smaller fragments are rich supplies for filter-feeding organisms such as mussels,

BELOW **A wide range of encrusting sponges, other invertebrate animals and even small fish live in the shelter of large brown seaweed holdfasts.**

barnacles, tunicates, anemones and polychaete worms. Planktonic feeders, such as copepods will also digest these smaller fragments of the algae. Detached seaweeds washed up on the shore are another valued resource. Lift up any drift weed that has been lying on a beach for a few hours and multitudes of sand hoppers will scramble for cover. The tide line of partially rotted vegetation is key to their survival.

Even larger organisms utilise seaweeds, including the herbivorous green turtles that eat both seaweeds and seagrasses. Some seabirds use floating mats of seaweed to rest upon and soft seaweeds make excellent nesting material. Most spectacularly, grey whales strip the invertebrates off giant kelps, by running the fronds through their baleen plates.

Mucus producers

If you have ever tried to walk amongst seaweed on a beach, you will be familiar with its slimy surface. When the seaweed is covered by the tide, the slime (consisting mainly of polysaccharides) dissolves into the surrounding seawater. Seaweeds also release sugars and amino acids into the water as a result of damage or due to regulation of their metabolism in response to environmental stress. These substances are referred to as dissolved organic matter (DOM) and it is thought that up to 30% of the carbon assimilated by *Laminaria* may be released in the form of DOM. The ecological importance of this large fraction is that bacteria and fungi and some protozoans feed on DOM. The seaweeds, therefore, play a role in the

ABOVE **Sea urchins are often seen grazing on large brown subtidal seaweeds such as *Laminaria* and *Macrocystis*.**

LEFT **Chitons grazing at the bases of holdfasts.**

regulation of bacterial activity within coastal waters. As the bacteria break down the DOM, nutrients such as nitrogen and phosphorus are released back into the water and can be taken up again by the seaweed for new growth.

Many of the organisms that feed directly or indirectly on the seaweeds live in close association with them. This is best demonstrated by subtidal forests of brown seaweeds, particularly the diverse faunas that live amongst the holdfasts of the kelps. These are often covered with filter-feeding sponges and anemones, and teaming with worms, crustaceans and brittlestars. Snails and sea slugs also graze on the fronds themselves, or more frequently, the biofilms of microalgae and bacteria that accumulate on the seaweed surfaces.

Seaweed as shelter

Inevitably, these assemblages of seaweeds and animals attract fish for the rich pickings they provide. The fish help to keep the numbers of grazing organisms down to reasonable levels. But it isn't just the food that attracts the fish since the dense growths of seaweed form an effective refuge from larger predators such as birds and dolphins. Shoals of plankton-eating fish hide amongst stands of seaweed, coming out of cover to feed only at certain times of the day. Some fish have an elaborate camouflage, such as the sargassum fish (*Histrio histrio*), an anglerfish, that is covered with fleshy and leafy appendages that resemble floating *Sargassum* in which it lives. Even more spectacular are the weedy and leafy sea dragons (*Phyllopteryx taeniolatus*

LEFT **The Sargassum fish (*Histrio histrio*) has elaborate camouflage to make it hard to see amongst the floating *Sargassum* in which it lives.**

RIGHT **Seaweeds can even form a significant portion of the material used by seabirds for their nests.**

and *Phycodurus eques* respectively) that shelter among the fronds of seaweed, their colourful, leafy flaps of skin blending in seamlessly. Giant kelpfish (*Heterostichus rostratus*) are found in kelp beds from British Columbia to California where their blade-shaped bodies help them to hide among the seaweeds. To further their disguise, they can change colour to match any variation in hue of the seaweed backdrop.

Giant seaweeds and sea otters

It is the association between the sea otters and the giant *Macocystis* kelp beds off the Californian and Alaskan coasts that is probably one of the best reported seaweed-based food webs. *Macrocystis* is harvested in vast quantities for the seaweed processing industries (see p. 85). The sea otters do not eat the seaweed, although they may anchor themselves in rafts of *Macrocystis* fronds on

the surface of the water when they wish to rest. The otters are interested in the kelp forests because of the invertebrates that are found there, in particular the sea urchins. Sea urchins are voracious grazers of seaweeds; a dense swarm can effectively strip an area of attached seaweeds. The damage they cause has been compared to the devastation left by a forest or bush fire. Most urchin species feed on fragments of seaweed that they catch with their tube feet but they can also feed on attached thalli. In field experiments where divers have regularly removed urchins, species of *Laminaria* were shown to increase in number and extend to deeper depths.

Along the coast of southern California dramatic reductions in *Macrocystis* beds have been recorded at various times during the past 200 years due to extensive grazing by sea urchins. Various horrific schemes have been initiated to remove the urchins and to encourage the re-establishment of the seaweeds. These ranged from burning the urchins with quicklime, which unfortunately also damaged many other marine organisms, to simply smashing the urchins with hammers.

The demise of the *Macrocystis* beds corresponded with a fall in the sea otter population in the area. However, in the 1960s otter numbers in the region began to increase and the kelp beds soon started to re-establish. It is thought that the feeding activities of the otters lowered the sea urchin population to the extent where the kelp beds were able to recover. Otters are certainly prolific hunters: 50 otters have been recorded eating 5000 red sea urchins, 300 mussels and 400 abalones in a two month period on the Monterey coast.

Is it really the otters?

Some researchers question the rise of the urchin populations being due solely to the demise of the sea otter population. Abalones compete with urchins for food and, when these were over fished in the 1930s, it is thought that the urchins were able to grow to densities that couldn't be controlled by other urchin predators such as lobsters, fish and starfish. The success of the urchins may have been partly due to their taking up DOM from local sewage outfalls. Bad storms can be highly destructive to urchins and cyclical climatic events such as El Niño may also play a role in the control of urchin populations. However, work done at other study sites such as islands in the Aleutian archipelago in Alaska has shown clear links between urchin increases and kelp bed decrease, and an 80% drop in the sea otter population between 1987 and 1997.

ABOVE **Sea otters feed on sea urchins and may therefore help prevent the urchins destroying underwater forests of the giant seaweed *Macrocystis*.**

Grazers, competition and defence

Wherever seaweeds grow they have to compete for space, inorganic nutrients and light. They are also in constant danger of being grazed on by gastropods, fish, sea slugs and sea urchins and have some sinister tactics for putting off predators.

Fighting for space

When there is mass settling and germination of spores of the same species in a confined area, competition naturally occurs between the thalli. In general this results in the individuals being smaller since there is an inverse relationship between density and size. This is particularly applicable to stands of fast-growing, short-lived species such as *Ulva*, *Bangia*, *Enteromorpha* and *Porphyra*. But even the much slower growing encrusting coralline species compete for space – they end up growing into and over each other, sometimes becoming quite indistinct as individuals.

The competition for space may be with another seaweed species, but it is not necessarily the larger species that is successful. For example, the large brown Californian *Egregia laevigata* is out competed by small red algae that propagate vegetatively to fill space as soon as it becomes free. The spreading reds prevent *Egregia* spores settling and it is only when the reds are removed by an external disturbance of some kind that spores of the larger species can settle.

LEFT **An 'army' of** *Littorina* **snails grazing green seaweeds from a rock surface.**

Clearing rock faces

Animals such as barnacles and mussels that cling to the rocks are also adversaries for space. Yet there are mechanisms by which seaweed inadvertently helps to strip the creatures from the rocks, so creating a space in which to grow. The robust *Postelsia* inhabits parts of the shore that are also occupied by barnacles and mussels. Small *Postelsia* have been seen growing firmly attached to barnacles. When they reach a certain size, the increased drag causes the seaweed, together with the attached barnacle, to be ripped from the rock surface. The bare area of rock is immediately colonised by *Postelsia* spores. *Postelsia* seldom grows on mussels, however, and can only invade mussel-covered rocks if the mussels are removed by some other means.

Another example of how the mechanical clearance of a rock face can influence where seaweeds settle is winter ice scour that clears rock faces in northern New England. *Fucus* and *Ascophyllum* species are able to settle and

BELOW **Growth of a diverse and extensive epiphyte flora on *Laminaria hyperborea* stipes.**

dominate ice-scoured shores, whereas on sheltered shores that are not prone to ice scour they are out competed by *Chondrus crispus*.

Hitching a ride

One of the most effective ways of gaining an advantage in the quest for space and light is to grow on the surface of a moving animal or on the surface of another seaweed. Organisms that do this are called epiphytes. Most seaweeds, from encrusting corallines to large membranous species, can grow as epiphytes, but very few seaweeds have to live this way. One is the wiry red *Polysiphonia lanosa* that is usually found growing as an epiphyte on *Ascophyllum* thalli. Most epiphytic species are fast growing, short-lived and filamentous, although encrusting and large membranous species also grow as epiphytes.

But it is not only seaweeds that grow as epiphytes; there are many microalgae, in particular diatoms (silica-encased unicellular microalgae), that grow as epiphytes on larger seaweed thalli. These attach themselves with mucus, sometimes forming layers many cells thick. Bacteria can also become associated with these biofilms, contributing their own slime to mucus layers and developing complex films. Other organisms, such as bryozoans and the spirorbid polychaete worms that live inside tubes glued to the surfaces of thalli, are also seen in large numbers, particularly on perennial brown seaweeds.

An epiphyte's size is ultimately determined by the size and strength of the host seaweed. This is because the epiphytes increase the drag on the host and if the growth of epiphytes becomes too massive, the host may become detached. Extensive epiphyte growths also attract invertebrates that further increase the drag and weight. Robust *Laminaria* or *Macrocystis* stipes can support dense growths of epiphytes, including large membranous red seaweeds like *Palmaria palmata*. The epiphytes on smaller *Fucus* thalli tend to be restricted to smaller filamentous or short-lived species.

Generally, epiphytes do not grow on rapidly growing parts of the thallus. Therefore *Laminaria* and *Macrocystis* stipes may have large standing crops of epiphytes whereas the rapidly growing fronds or blades are largely free of epiphytes. Fast-growing species such as *Ulva* and *Porphyra* are rarely covered with epiphytes – they simply grow too fast for the epiphytes to take hold. The increased fluctuations in pH at the surfaces of rapidly photosynthesising tissues may also play a role in deterring initial settlement of the epiphytes.

Skin shedding

Epiphytes are not particularly welcome guests since they can harm their hosts by blocking out the light and taking up nutrients from the surrounding water. Layers of epiphytes can

RIGHT **Tuft of a brown filamentous *Elachista scutulata* growing as an epiphyte on the surface of the large brown seaweed *Himanthalia elongata*.**

OPPOSITE Sometimes an epiphyte can become bigger than the host on which it is growing. In this case the red *Palmaria palmata* swamps its *Fucus serratus* host.

RIGHT The brown seaweed *Himanthalia elongata* sheds outer layers of the cell wall to remove epiphytes growing on the surface. The imprints of the cell walls are seen on the flap of 'skin' being lifted off the new layer below.

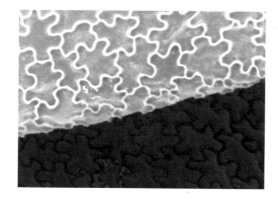

BELOW Snails retreat into these holes at low tide, coming out when covered by water to graze a very defined patch of green seaweeds around the holes.

act as effective barriers to exchange processes, including gaseous exchange, reducing the host's growth rate. Amphipod and other grazing animals may help to keep epiphytic growth down. However, some species, including *Chondrus crispus*, *Halidrys siliquosa* and *Himanthalia elongata*, regularly slough their surfaces, thereby ridding themselves of any build up of epiphytes. *Enteromorpha intestinalis* constantly produces new cell wall layers, while shedding

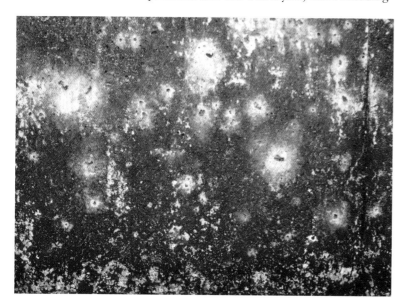

outer layers of the wall, to ensure that it remains epiphyte free. Another mechanism employed by some species is to produce chemical deterrents to prevent epiphytes settling. Phenolics and halogenated lipids are thought to play such a role, but numerous other chemicals have been implicated.

Trying not to be eaten

A wide range of herbivores, including fish, urchins, molluscs, amphipods, copepods and polychaete worms, eat seaweeds. As demonstrated by the effect of the urchins on the Californian kelp beds (see p. 57), grazing pressure can be tremendous, resulting in the collapse of seaweed populations.

As a defence, seaweeds have developed an arsenal of chemicals, including phenols, terpines and alkaloids. However, the production of such compounds is energetically expensive and seaweeds that have large stores of defensive anti-grazing chemicals tend to have slower growth rates than species without them. Rather than maintaining high concentrations of grazing deterrents all the time, some *Fucus* species produce greater quantities of phenolic compounds only when grazers are present. The red *Laurencia*, brown *Dictyota* and the tropical *Caulerpa* all produce toxins that are effective in deterring grazers.

Another effective anti-grazing strategy is shown by brown *Desmarestia* species, which actually stores sulphuric acid in vacuoles inside its cells. Any grazer that penetrates the thallus will get a swift dose of acid. The acid is so potent that when collecting seaweeds, it is important to separate *Desmarestia* specimens from other species since the acid effectively kills them. In fact, if damaged, the thalli themselves quickly disintegrate. Sea urchins that were kept on a diet of *Desmarestia* had very eroded teeth due to acid damage.

Many species do not just rely on the production of noxious anti-grazing agents, but also have unpalatable structural defences. The tropical brown *Turbinaria* is a good example, being characterised by a tough spiny thallus. Calcified species like *Halimeda* and *Corallina* are certainly less palatable to grazers. The opposite is true of fast-growing, short-lived species such as *Ulva* and *Enteromorpha*. These seaweeds do not produce anti-grazing chemicals or tough tissues. They simply rely on rapid growth rates in order to replenish tissues lost to grazers.

LEFT *Desmarestia* contains sulphuric acid within its cells, a very potent anti-grazing agent.

LEFT The tough spiny fronds of the tropical *Turbinaria* make it unpalatable to many grazers.

Herbivores are not all bad

Though seaweeds are food for a whole host of herbivores, in general less than 10% of biomass tends to be actively grazed. Many of the molluscs that appear to be grazing on the seaweeds are actually eating the films of epiphytic microalgae and bacteria that grow on the surfaces of the seaweed.

Even in the kelp forests, much of the damage done by urchins is not due to the urchins eating all the forest. It is due to the way the urchins weaken the stipes or holdfasts so that the thalli break off and are lost. Moderate grazing pressure can actually increase the diversity of algal species, by preventing the dominance of large canopy species or fast-growing species that may take over if not kept under control.

'House building' amphipods and 'farming' fish

Some surprising associations have developed between seaweeds and grazing animals. Some species of sea slug, for example, graze specifically on seaweeds that contain agents for deterring fish. By storing the anti-grazing agent in its own tissues the sea slug then has a deterrent to stop itself being predated upon by fish. A Caribbean amphipod, *Pseudamphithoides* constructs enclosures from *Dictyota* thalli, encouraged by chemicals released by the *Dictyota* itself. The chemicals are released to deter fish grazing, and the enclosures therefore also protect the amphipods from predatory fish. There are even examples of 'farming fish' (species of tropical damselfish) that maintain specific

LEFT **Limpets carrying tufts of *Enteromorpha* on their shells. This is an ideal place for the seaweeds to grow since the limpets cannot turn round to graze these epiphytes.**

stands of seaweed by careful selective grazing. Sometimes they maintain 'pastures' of just one species or a limited number of species over large expanses of reefs.

The effect of grazing and competition on zonation

It is a commonly held view that the upper limits of a particular species on the shore are set by the tolerances to one or more of the physical stresses mentioned previously, while the lower limits are set by the effect of grazing and competition. However, when grazers are excluded the upper distribution of some seaweed species can extend upwards. In the subtidal zone light is the obvious single most dominant factor in determining the distribution of seaweeds and zonation patterns. The intertidal region is somewhat more complicated since none of the physical 'stress factors' act in isolation. Rather than being a response to one particular stress, distribution limits result from the interaction of several factors prevailing at a particular place on a shore. Population differences in tolerances within a species, even between

BELOW **The distribution of seaweeds on the shore is a combination of both biological interactions and the physical forces that are unique to any one particular shore. Pacific Coast, Chile.**

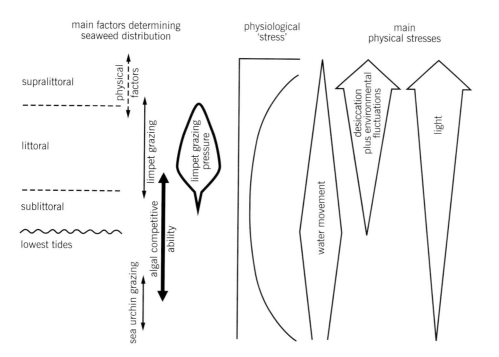

main factors determining seaweed distribution

physiological 'stress'

main physical stresses

supralittoral

littoral

sublittoral

lowest tides

physical factors

limpet grazing

limpet grazing pressure

algal competitive ability

sea urchin grazing

water movement

desiccation plus environmental fluctuations

light

LEFT **Summary of the dominant factors governing the distribution of seaweeds on a shore, where limpets and sea urchins are the main grazers of seaweeds.**

populations on the same shore, make it difficult to discern clear-cut trends. Seasonal patterns add yet further complexity.

It is deceptively easy to measure the response of a collection of seaweed species to a stress factor and relate the findings to where species are found on the shore. However, until such responses are investigated over the whole range of temperatures, light conditions (including darkness) and salinities that may be experienced, single factor experiments at best give us only a hint of the real situation. Explanations for zonation should not be

sought in resistance to stress of mature seaweeds. The successful colonisation of a particular shore depends on the recruitment of microscopic stages, and so different life cycle stages must be considered in any comprehensive studies.

A more accurate interpretation of the causes of zonation is probably that the 'potential' limits of the distributions of a species are set by physical and chemical limitations. The actual distribution on a shore is finely tuned by biological interactions such as grazing pressure and competition.

Seaweeds at the extremes

Seaweeds are ubiquitous, being found in polar, tropical and temperate waters around the globe. They are found in regions where tidal activity generates long periods of emersion and in places where there is hardly any tidal amplitude at all. Some species have an almost global distribution, whereas others are limited to small geographical regions. Several specialised seaweed communities are distinguishable and are worth brief discussion here.

The Baltic Sea

One of the most curious 'marine' communities is that found in the Baltic Sea where seaweeds such as *Fucus vesiculosus* and *Chorda filum* grow alongside freshwater flowering plants such as *Ranunculus*, *Myriophyllum* and *Potamogeton*. Although the Baltic is connected to the north Atlantic, the seawater in this land-locked sea is very dilute. The salinity ranges from 8 at its mouth to less

BELOW **In the Baltic, seaweeds typical of marine sites such as the brown *Fucus vesiculosus* grow alongside species more commonly found in freshwaters, such as the dense swirls of *Cladophora glomerata*.**

than 2 in central parts of the sea. The Baltic Sea is often covered in ice for several months in the winter, but it is evident that salinity plays the major role in determining the distribution of Baltic seaweeds. The large brown *Fucus serratus* and *F. spiralis* do not penetrate much further into the Baltic than the mouth where salinities are between 7 and 8, whereas *F. vesiculosus* is able to grow in salinities of 4 (characteristic of waters far into the inner parts of the Baltic Sea).

Seaweeds found growing in the Baltic have different salinity tolerances to populations of the same species from truly marine sites. The Baltic seaweeds are significantly more tolerant of low salinity waters, but have a reduced tolerance to salinities above seawater concentrations when compared with their marine counterparts.

The Baltic also contains normally freshwater species, such as *Cladophora glomerata*, that in places grows alongside *C. rupestris*, which is of definite marine origin. Freshwater populations of *C. glomerata* do not tolerate any salt in the water but in Baltic waters the seaweed is found growing in salinities up to 6, and can tolerate salinities up to 16.

Baltic seaweeds close to their salinity limits become more dependent on asexual and vegetative reproduction. When species that are normally marine grow at low salinities individuals generally become much smaller compared to those living in full seawater. This is not just the gross thallus size, but also the sizes of cells, and may even include structural differences. *Chorda filum* from marine sites has thin continuous diaphragms traversing

the hollow bootlace-like thalli. In the Baltic form the diaphragms are not continuous, but rather complex dissected structures. These smaller forms are often referred to as being 'depauperate', implying some sort of inferior, rather feeble form. However, this is clearly not the case and despite the size differences, Baltic populations are just as efficient in photosynthesis and growth as their marine counterparts.

Seaweed evolution in the Baltic

It is thought that seaweeds started to grow in the Baltic about 7500 years ago when seawater from the North Sea began to flow into the freshwater basin, making it brackish, and that subsequent natural selection has resulted in the considerable changes observed in the species found there. However, the Baltic seaweeds have not been isolated long enough for speciation to take place, and the Baltic populations may be best considered to be subspecies of those from marine sites. The

ABOVE **Throwing a hook from a Baltic small rocky island to sample the** *Fucus vesiculosus* **population that is permanently submerged in Baltic waters, in contrast to populations outside of the Baltic that are mostly exposed at low tide.**

relatively recent isolation of the Baltic Sea and the intriguing flora that has developed has led some to refer to the Baltic as being a 'cradle of plant evolution'. This seems an appropriate label and is one of the few places where both marine and freshwater specialists have the chance to join forces in unravelling the complexities of evolution.

The high shoreline of the Baltic is often scoured by winter pack ice. The ice can be very effective at removing any attached seaweeds, resulting in the ice-scoured part of the shore being dominated by fast-growing, short-lived seaweeds such as *Cladophora*, *Ectocarpus* and *Enteromorpha*.

ABOVE **The upper zones of Baltic small rocky islands are often covered in dense growths of short-lived species such as *Cladophora* that are rapidly scoured from the rocks by ice in winter.**

LEFT **Blocks of ice scraped across rock surfaces by moving water are very effective for clearing all seaweeds from the surfaces.**

Seaweeds at the poles

The annual freezing over of coastal waters, consolidation of pack ice, and its subsequent melt is a significant feature of the Arctic and Antarctic. Some seaweed species, including the red *Iridaea cordata* in the Antarctic, occur at latitudes of 77°S, where they experience up to ten months of ice cover. These waters rarely go above 5°C (41°F) and for much of the year they are closer to the freezing point of seawater at −1.8°C (29°F). In recent years growth characteristics of several Arctic and Antarctic seaweeds have been studied and found to be restricted to low temperatures. However, this does not mean all seaweeds found in these waters are restricted to these low temperatures. In fact only about 5% of the seaweeds found in Arctic waters are endemic (only found there) to the Arctic. In the Antarctic a greater proportion of the species found there are endemic (30%). The ability of Antarctic and Arctic seaweeds to photosynthesise and grow at 0°C (32°F) at rates similar to temperate species at higher temperatures certainly reflects a considerable adaptation to low temperature.

Modified enzymes that work at low temperatures bring about efficient metabolism. Another major factor is the composition of the cell membranes, which consist mostly of different types of lipids (fats), and have to remain fluid for normal cell processes to take place. As temperatures decrease, polar seaweeds alter the composition of lipids in their membranes to stop them freezing.

Growing in the poor light of the poles

Another characteristic of polar regions is the very seasonal light conditions: day lengths of up to 24 hours in the summer and long periods of darkness in winter. Polar seaweeds are adapted to low light conditions. Several species, including the brown *Laminaria solidungula* from the Arctic and the red *Palmaria decipiens* from the Antarctic, actually begin to grow during periods of darkness at the end of the winter, when still covered by the ice. These seaweeds haven't found a way of growing without light – they grow using the starches and other metabolites that they stored in the previous year's growth period. The new tissues they produce are

RIGHT **In shallow Arctic waters, seaweeds can become entrapped within the ice that forms every winter.**

LEFT **Healthy** *Laminaria* **and** *Fucus* **retrieved from a depth of 5 m (15 ft) under the winter pack ice of the White Sea, Russia.**

ready to begin photosynthesis as soon as light becomes available when the ice breaks up. This kick-start maximises the growth period during the short summer months. The development of new blades in the dark is probably controlled by circannual rhythms, a hypothesis supported by the fact that when these seaweeds are grown under constant conditions they still show the same rhythms in growth patterns.

Seaweeds in the tropics

At the other extreme, seaweeds are an important component of tropical marine environments. The geographic boundaries of tropical seaweed species sometimes might coincide with those of reef-forming corals that are found between the 22° north and south isotherms. Generally it is thought that the tropics host the greatest biodiversity of marine organisms. Rather surprisingly the greatest diversity of seaweed species does not occur in tropical waters, possibly because in tropical waters the corals effectively out compete the seaweeds for space. Instead, the greatest diversity of seaweeds is found along warm temperate coasts such as Japan, the Mediterranean and notably the southern coast of Australia. Approximately 1100 seaweed species grow in these Australian waters and over 25% of these species do not grow anywhere else in the world. One of the reasons for this is because the coastline is particularly long with a relatively constant water temperature. Transfer and establishment of viable seaweed populations are therefore easy along the whole length of the coastline. There is also a great diversity of shore types, enhancing the development of the diverse seaweed flora. Other similarly long coastlines in the world are associated with large gradients of temperature because they run north-south.

Clear waters

Tropical waters, in which temperatures reach over 30°C (86°F), are characteristically clear so light can penetrate to great depths. The deepest recorded seaweed species are found in such clear waters, living on a seamount off San Salvador Island in the Bahamas. For light to penetrate to the record 268 m (879 ft) the water has to be devoid of practically all particulate matter, including microscopic plankton. The waters of the tropics have low concentrations of inorganic nutrients, and

LEFT **Coral reefs are not just a habitat for corals, but also a wide range of seaweed species.**

therefore do not support the rich soups of phytoplankton that quickly grow in temperate waters and which subsequently prevent the light from reaching similar depths.

Sandy lagoon flats

In near-shore sand and lagoon flats, including areas where seagrasses grow, species such as *Caulerpa, Udotea, Padina, Halimeda* and *Penicillus* are frequently found. Some of these, like *Caulerpa* and *Penicillus*, spread by subsurface runners that are anchored by rhizoids. These help to stabilise the sediment reducing the effects of erosion, as well as ensuring the seaweeds are not easily dislodged. Many of these species, a good example being *Halimeda*, are calcifying and fragments can form a high proportion of the sediments in tropical regions. A limited number of species, including *Udotea* and

Penicillus, can move nutrients from the pores of the sediments through their rhizoids.

Seaweeds and corals

On coral reefs there often doesn't appear to be dense seaweed growths. One major reason for this is that corals themselves out compete the seaweeds for space. Another reason is that effective grazing by herbivores (fish and urchins) tends to keep the growth of seaweeds on reefs to a minimum. However, species with effective anti-grazing defence strategies such as *Turbinaria* do at times reach high populations. As soon as controlling factors such as herbivores are removed, the density of seaweeds on a reef increases dramatically, shading the underlying reef from incident light.

Coral reefs are generally formed in low nutrient waters. If the nutrient content of the water is increased, due to seasonal events

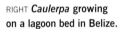
RIGHT *Caulerpa* growing on a lagoon bed in Belize.

ABOVE **Calcareous red seaweeds such as** *Porolithon* **help to cement corals together in tropical reefs, the seaweeds contributing to over 70% of the reef in some cases.**

such as mixing of nutrient-rich bottom waters into nutrient-poor surface waters, or the release of sewage or agricultural runoff, the growth rates of seaweeds are stimulated and they overgrow the reef. This is normally a short-term event, until the elevated nutrients are exhausted. However, after a sustained period of increased nutrient loading, the growth of algae may be enough to completely overshadow the corals below, thus killing them. In regions of tropical coasts where upwelling of nutrient-rich waters occurs on a regular basis, corals do not develop well and these regions are characterised by seaweeds. The coast of Oman is a good example where northern shores are characterised by having well-developed coral reefs, while in the south (where upwelling occurs) there are large kelp beds and no corals.

Calcifying seaweeds and coral reefs

Calcareous seaweeds are major contributors to the structure of coral reefs: they lay down

so much calcium carbonate in their thalli that they can form a major portion of the calcium carbonate that cements reefs together. In fact, some reefs can be predominantly calcifying algae, cementing corals together that may only make up 30% of the reef. Species such as *Porolithon* and *Lithophyllum* are typical of the algal reef crests or ridges that form at the wave-exposed, seaward part of a reef. These particular algae, although they grow slowly, are able to withstand considerable drying out, sand scour and wave action, and are not eaten by herbivores.

Despite the fundamental importance of calcifying seaweeds for the formation of reef systems in the tropics and subtropics, there are several blue green algae and green filamentous algae that actually bore into the corals. These penetrating seaweeds play a significant role in the erosion of reef structures, even eventually leading to the breakdown in reef structure.

Coral bleaching

The structure of coral reefs is clearly in a delicate balance: nutrient loading and the degree of grazing activity combine to determine whether seaweed or corals dominate. Another factor is temperature. Even small shifts of one or two degrees in temperature can lead to substantial coral bleaching and death of corals, especially when the rise in temperature is associated with influxes of low salinity water and low light levels. Coral bleaching may also be enhanced by viral infections. When coral bleaching occurs to any large extent the increased space becomes available for seaweeds to settle.

ABOVE **Dead or damaged corals following 'bleaching events' quickly become smothered in dense growths of filamentous green seaweeds.**

RIGHT **Exposed mangrove aerial roots are often covered by small seaweed species that form layers cemented together with mud.**

Arabian Gulf and the Caribbean. At some of these sites up to 70% mortality of corals were recorded, and massive growths of seaweed were observed on the dead corals.

Seaweeds in mangroves

Mangrove swamps are a well-known feature of intertidal coastal regions throughout the tropics and subtropics. Mangroves can form thick forests characterised by dense root systems that anchor the plants into thick muddy sediments. The root systems can form almost impenetrable masses that make mangrove forests important refuges for many types of animal, ranging from pelagic fish to crabs and mudskippers.

Several global coral bleaching events were reported in the late 1990s to 2001. In 1998 massive coral bleaching was experienced in at least 60 countries and island nations in the Pacific Ocean, Indian Ocean, Red Sea,

The aerial roots of mangrove trees, which are submerged periodically by rising tides, are colonised by a wide range of sponges, barnacles, oysters and seaweeds.

LEFT **Mangrove roots in an enclosed lagoon, at Calabash Cay, Belize, provide ideal structures for a diversity of epiphytic seaweeds to grow on.**

The seaweeds that grow on mangroves tend to be small, turf-like species of *Caloglossa*, *Catenella* and *Bostrychia*. These are all highly tolerant of changes in salinity and drying out, which enables them to survive periods of exposure that are particularly harsh in such regions. However, because mangroves are such muddy places, the seaweeds are often caked in thick layers of slow-drying mud at low tide. The mud offers protection against drying out but interferes with photosynthesis.

Other types of seaweed grow attached to mangroves on the roots and lower parts of the trunks, although the high turbidity of the water limits the number. Any venture into a mangrove forest quickly reveals that a lot of drifting material is trapped by the root systems and this applies to seaweeds as well. Sheltered mangrove systems can become clogged up with unattached seaweeds, which eventually die, providing a rich source of food for detritus feeders such as the fiddler crabs and mudskippers. Of course, the trapped seaweeds and the epiphyte turfs are important food sources for amphipods and isopods that feed off the seaweeds directly.

Problem seaweeds

Dense mats of *Enteromorpha*, *Ulva*, *Cladophora* and other fast-growing seaweed species that periodically clog up sheltered bays, inlets, waterways or shallow seas are commonly reported at sites from all regions of the world. Sometimes they simply get in the way of swimmers and boats and they are often washed up on to shores where they lay rotting in unsightly, foul-smelling masses. Under natural conditions seaweed growth is limited by light, temperature and nutrients. Given the correct conditions and copious supplies of nitrogen, phosphorus and other trace elements, seaweeds can achieve phenomenal growth rates.

Eutrophication

Despite 20 years or more of tight legislation and control, it is still common in many coastal sites around the world for agricultural waste and domestic sewage to be dumped straight into rivers and coastal waters. Such material, even when treated, is often high in nitrogen and/or phosphorus and it stimulates the growth of both phytoplankton and seaweed. The accumulation of nutrients in

LEFT **The consequences of eutrophication: masses of rotting *Ulva* and *Enteromorpha* have been washed up on to a shore.**

waterways is called eutrophication. It must be stressed, however, that in most coastal waters even highly concentrated pulses of nutrient-rich water are generally effectively diluted. But where water exchange is limited and dilution inhibited the increased nutrients can be large enough to stimulate a problematic growth of algae.

Knock on effects of eutrophication

It is not just the seaweed biomass on the surface of the water that is the problem. When the algae die they generally sink to the sea floor where they decompose. The decomposition by bacteria and fungi consumes oxygen. Animals living on the bottom of the sea and within the sediments require oxygen to live and if oxygen is exhausted these animals die. In the worst cases eutrophication leads to the death of masses of fish. Even mudflat and saltmarsh areas can be overgrown by such algae, the dense mats effectively cutting off the oxygen supply to the sediments below.

Even before such drastic changes take place, eutrophication is signalled by changes in species composition. Again it seems to be the fast-growing, short-lived seaweeds that gain a competitive advantage over slower growing species. Eutrophication increases the rate of phytoplankton growth and the increased phytoplankton biomass in the water can dramatically alter the quality of underwater light fields as well as reducing the depth to which light penetrates. Eutrophic waters also stimulate the growth of epiphytic algae on the surfaces of larger seaweed species, again effectively reducing the light available to the host seaweed. The eutrophication and increased phytoplankton concentrations of Baltic waters since the 1940s is thought to account for the fact that *Fucus vesiculosus* populations are now found only at depths above 8 m (26 ft), whereas they used to be seen at depths of 12 m (39 ft).

Of course, any dense growth of seaweed that does accumulate can be collected from shores where it washes up or even harvested directly from the water itself. It can then be used as fertiliser, animal fodder or soil conditioner.

Alien seaweeds

The ever-increasing level of marine traffic is causing species of seaweeds to be transported large distances from where they originated. Although seaweeds may be transported on the hulls of ships, this vector is unlikely to be a highly effective means of transport. The dumping of large quantities of ballast water is the more likely means by which long-distance transport takes place. However, 'alien' seaweed species have also been introduced in

RIGHT **Inspecting** ***Sargassum muticum***.

more curious ways, such as on the shells of oysters and other introduced shellfish and in the packaging of seafood. They have also been deliberately introduced for aquaculture purposes.

The majority of introductions of non-native species inevitably die since living conditions are not optimal for their survival. For example, it is unlikely that an Antarctic species will survive when released into tropical waters. Even if a seaweed species does survive, its population is likely to be very small. However, there are several examples of introduced seaweeds that have taken hold with dramatically adverse effects on the ecosystems they have arrived in. On occasion the consequences for local economies have been disastrous.

Introduced *Sargassum muticum*

The brown seaweed *Sargassum muticum* has a range stretching from British Columbia to Baja in California. It is also found along the coasts of Britain, France, Scandinavia and the Iberian Peninsula. However, the species originated in Japan and is thought to have gained its worldwide distribution through being transported with Japanese oysters. The species is particularly tenacious, with fast growth rates, high reproductive rates and the ability to spread vegetatively. Its high reproductive rates are enhanced by it being monoecious and self-fertile. It can grow in wave-exposed sites and in water up to 10 m (33 ft) deep. Its rate of expansion along coastlines has been estimated at hundreds of kilometres per year and it can out compete native seaweed communities effectively.

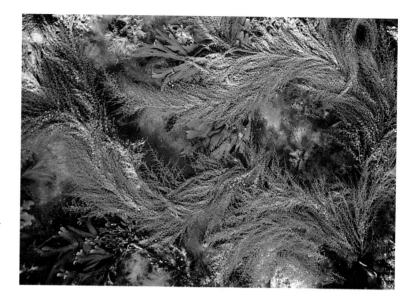

Because of its prolific growth it has become a nuisance alga, forming detached drifting mats and clogging marinas and recreational areas.

Introduced *Undaria pinnatifida*

Undaria pinnatifida is a large kelp native to Japan, Korea and China, where it is extensively cultivated as a food (wakame). Its life cycle is similar to that described for *Laminaria*, with a large sporophyte and microscopic gametophyte stages. Populations of *U. pinnatifida* are established throughout Europe, New Zealand Argentina and Australia. It is likely that the initial introduction into Europe was accidental together with young Pacific oysters. However, subsequent introductions were intentional, with attempted cultivation of the species in the Mediterranean and on the Atlantic coast. Its introduction from France to Britain is thought to have been via coastal boat traffic. In New Zealand, the initial introductions are

ABOVE **The worldwide distribution of** *Sargassum muticum* **is thought to have been achieved due to transport with Japanese oysters.**

RIGHT *Caulerpa taxifolia* clone invading seagrass bed.

thought to have been via shipping from Asia as gametophytes in ballast tanks or as sporophytes attached to the hull of a vessel.

Like *Sargassum muticum*, *U. pinnatifida* has a competitive edge over many species in the regions to which it has been introduced. It is able to colonise a range of shores of varying wave exposure and can grow at depths of up to 15 m (49 ft). *U. pinnatifida* rapidly colonises new or disturbed sites, including navigation buoys. A survey carried out in selected New Zealand ports found that 16% of the fishing vessels and 40% of yachts and launches inspected, were fouled with *U. pinnatifida*. It only takes one of these vessels to visit a non-infested area to establish a new *U. pinnatifida* population. Even diving gear could be a possible vector of the species.

The killer alga *Caulerpa taxifolia*

One of the most widespread reported 'invasions by alien algae' of recent years has been the rampant extension of the green alga *Caulerpa taxifolia* through the Mediterranean Sea. The alga has won such a notorious reputation that it is frequently referred to in the press as the 'killer alga'. It is thought that *C. taxifolia*, which is a native of the tropics, was accidentally released in the early 1980s in front of the Oceanographic Museum of Monaco, where it was grown because it is in fact an exceptionally beautiful vivid green alga.

It now grows along the coastline of six Mediterranean countries, covering huge areas from shallow waters to depths of up to 100 m (328 ft). Although the strain does not reproduce sexually at the 'low' temperatures of the Mediterranean, it multiplies extremely efficiently by vegetative growth and fragmentation. Yacht anchors and fishing gear transport it from one harbour to another. It is toxic to herbivores and so is not kept in check by fish or urchins. The toxins and the phenomenal growth rates of *C. taxifolia* have resulted in the expansion going largely unchecked with the subsequent demise of local seagrass and seaweed beds. The seabed very quickly becomes a dense forest of nothing but *C. taxifolia*, which is not only a disaster for seagrass and algal biodiversity, but also wipes out all the animals that relied on those seaweeds and seagrasses for food and shelter.

Recently populations of non-native *Caulerpa*, evidently the same as the Mediterranean species, have been identified in waters off southern California and Australia. Again, it is thought that these outbreaks originated from accidental introductions from aquariums and they are growing with an equal vigour to the Mediterranean populations.

The transport, storage, and release of *C. taxifolia* are now illegal in many of the countries where outbreaks of the alga have occurred, but the race is on to try to find methods to eradicate the seaweed. Over the past few years a whole range of ambitious methods for getting rid of the alga have been proposed and tried: vacuuming the algae up, applying dry ice, ultrasound, hot water jets or toxic doses of copper and salt. Some sea slugs are known to feed on *C. taxifolia* and researchers have suggested that these could be introduced as a biological control. To date nothing has been successful or practical enough to allow widespread control. The accidental release of *C. taxifolia* has been a costly one that whole ecosystems may take decades to recover from when and if a mechanism for controlling its expansion can be found.

Uses of seaweed

The Roman poet Virgil is quoted as saying that "there is nothing more vile than seaweed", and looking at (and smelling) the masses of rotting weed that accumulate on beaches after a storm it is easy to see where the sentiment stems from. However, for thousands of years seaweed has been used for its potent healing properties, even supposedly helping to rejuvenate lost looks. Seaweeds are effective fertilisers, soil conditioners and are a source of livestock feed. They are used in a huge range of products from ice cream to shaving foams and fabric dyes. Seaweed has long been a source of human food. In Japan, for instance, it constitutes up to 10% of the average diet. Today seaweed is served up in the trendiest of restaurants throughout the world.

Seaweed: a panacea for all

There seems to be no end of seaweed-filled lotions and potions that promise to protect your skin from harmful pollutants, smooth away wrinkles, control cellulite and even help you to lose weight. Shampoos, bath gels and soaps fortified with the invigorating

LEFT **Collecting *Fucus* species on a Brittany shore.**

properties of the sea in the form of seaweed extracts, or even dried seaweed, are commonplace. Eating seaweed is also heralded as effective against heart disease, water retention and rheumatism, while stabilising blood sugar levels. Seaweed is also a superb source of minerals and vitamins. These claims are not new: ancient Japanese and Chinese manuscripts describe the collection of seaweeds and their use as medicines and food. Roman women used rouge prepared from *Fucus* species, and today there is a cosmetic cream whose main active ingredient is 'Fresh Pacific Ocean Kelp' and sells at over US$75 (£53) for a few teaspoons.

Gums and gels from seaweeds

Much of the seaweed industry (worth over US$550 million (£390 million) per year) is involved with the extraction of alginates and gums. These products are superb emulsifiers, gelling agents and thickeners, and seaweed-derived gums are found in pudding mixes, ice creams, toothpaste, medicinal creams, sauces and drinks. The seaweed gelling agents can be broken down into three major classes: agar, carrageenans and alginates.

Agar is derived from red seaweeds, such as *Gelidium, Gracilaria, Hypnea* and *Pterocladia*, that are harvested in many countries around the world including Chile, India, Mexico, California, South Africa and Japan. Currently, there is a world shortage in agar and it commands a high price in international markets. It is used most famously as a microbiological growth medium but is also used in the food industry.

Carrageenans are extracted from a different group of red algae including *Chondrus* and *Gigartina*. The bulk of the carrageenan supply comes from seaweeds harvested from wild populations in Canada.

Alginates are found in the cell walls of many of the larger brown seaweeds. The main producers of alginates are the USA, Norway, China, Canada, France and Japan, where species such as *Macrocystis, Ascophyllum* and *Laminaria* are harvested. The product ends up in a range of goods, from wound dressings to vivid fabric dyes. One company alone harvests over 250,000 tonnes (240,605 tons) a year of wild *Ascophyllum nodosum* and *Laminaria digitata* from around the coasts of Canada and Norway. Great efforts are taken to reduce the environmental impact of these operations, by limiting the frequency of harvesting and making sure that the seaweeds are cut so that the stumps are left behind to regenerate.

ABOVE **Seaweeds and seaweed extracts end up in a large variety of products that are in everyday use.**

Nori production

Harvesting from the wild cannot satiate the world demand for certain species. A good example is nori or *Porphyra*, which in the UK is eaten as laverbread. In Japan over 60,000 hectares (148,480 acres) of Japanese coastal waters are given over to producing an annual crop of about 350,000 tonnes (344,470 tons), worth over a billion dollars. China produces about one third of this, making this somewhat unimpressive seaweed the single most valuable crop grown by cultivation in the sea. The nori industry was revolutionised by the work of Kathleen Drew who discovered the *Conchocelis* phase in the life cycle of *Porphyra*. Nori farmers now seed the nets that are used to grow commercial *Porphyra* from choncospores released from the *Conchocelis* phase (see p. 25). However, it is not just *Porphyra* that has a huge industry geared to its cultivation and processing; *Laminaria* (kombu), and *Undaria* (wakame) are also cultivated in carefully controlled conditions and grown in colossal coastal farms where the growth and yield are optimised.

Seaweeds as animal feed

One of the earliest uses of seaweeds was probably for feeding domestic animals, either by letting the animals graze on the shore at low tides, or by collecting the seaweed to make into feeds. Ronaldsay sheep on the

RIGHT **Harvesting nori (*Porphyra*) that grows suspended from the nets being pulled into the boat.**

RIGHT **Pile of *Laminaria* collected in Ireland for processing.**

northernmost Orkney Island, off the coast of Scotland, have a staple diet of seaweed. The sheep are confined to the foreshore by a drystone wall (the sheep-dyke) that runs around the island. During the lambing period the ewes are brought inside the dyke to feed on grass for three to four months before being returned to the shore. There are also many reports of undomesticated animals resorting to a diet of seaweed at times, including polar bears, rabbits, arctic foxes and wild deer.

Fertilisers

The use of seaweeds in agriculture has not just been confined to animal feeds. Seaweeds have long been used as fertilisers. Throughout history farmers living within access to the coast have collected drift weed as well as picking seaweed from the shore to use as soil conditioners and mulches. This still goes on today, although a much greater industry is the production of liquid fertilisers from seaweed extracts (mostly dried brown seaweeds). Another commonly used seaweed-derived soil conditioner is maerl, the heavily calcified red algal species that grows in offshore beds. This is dredged from the sea floor and normally crushed to a powder before being sold.

The benefits of seaweed-derived fertilisers and soil conditioners are well documented, although the commonly cited fact that they contain valuable stores of trace elements may have been over stressed. They are often applied in such dilute forms that the amount of trace elements derived from the seaweeds would be negligible. However, they have the advantage that they are free from terrestrial plant pathogens and fungi.

Biological scrubbers

Some seaweeds have exceptionally fast growth rates when light and nutrient supply is abundant. These are increasingly being harnessed as 'biological scrubbers' to clear effluent waters of a range of substances such as heavy metals or even high loadings of nitrates and phosphates. An example is the use of beds of *Ulva* to strip the nutrients from the very enriched waters being pumped from intensive fish farms, before the water is returned to the open sea. The *Ulva* is harvested when it reaches maturity, all the nutrients having been converted into seaweed. The seaweed can then be used as a fertiliser or soil conditioner. Where seaweeds are used as scrubbers to remove heavy metals, they concentrate the metals in a form that can be disposed of easily on harvesting.

The future for seaweed

There is currently an interest in identifying 'useful' products and chemicals from marine organisms, including the seaweeds. The most

BELOW **Diver swimming under ropes laden with** *Laminaria* **fronds on a seaweed farm.**

state-of-the-art analytical techniques are used to isolate compounds that may be beneficial to humans as new drugs, antibiotics, cancer treatments and so on. Mankind has used seaweeds since at least the first records were made. Who knows what we will find when we analyse them some more. It is certain, however, that we'll continue to eat seaweeds, use them in medicine, feed them to our animals and add them to our crops for some time to come.

Conservation of seaweeds

Our use of seaweeds has significant implications for the coastal zones in which they grow. The problems of introduced species and the disastrous effects these may have on natural seaweed populations were discussed above. The effects of rapidly growing human populations on the coast, as well as increasing trade, pollution and global climate change also threaten seaweeds. All these factors will impact upon the seaweeds and their ecosystems.

In particular, harvesting large quantities of wild seaweed is of concern and may have significant local ecological effects leading to reductions of biodiversity. No matter how it is done the cutting and removal of large canopy species disturbs the associated communities of animals and other seaweeds, as well as the fish and invertebrates that shelter and graze within stands of seaweeds. Even the dredging of maerl beds must have disastrous effects for the diverse communities of animals that are often found associated with them. This is compounded by the fact that the seaweed species that make the maerl beds are so slow growing, so recovery after dredging is slow.

Naturally the farming of seaweeds will go a long way to addressing many of these issues. However, as is clear from the experience of many aquatic farming systems, care has to be taken when choosing a location for such large-scale ventures. Any intensive system will dramatically alter local ecosystems, especially if nutrients and other factors are added to waters and localised water currents are affected. To date the only viable large-scale seaweed cultivation has taken place in Asia where the main demand is for seaweeds as food. The methods used are still highly labour intensive. The current reality of the seaweed market potential make the setting up of viable commercial seaweed farms elsewhere too risky, especially when compared to the relative ease of harvesting wild populations.

Removing seaweed may result in increased shore erosion since large stands of algae absorb and dissipate wave energy, thereby increasing natural sea defences. Even when intertidal seaweeds are removed the shore is more exposed and boulders and shingle may be displaced. Possibly the greatest problems are faced by coastal developments causing increased run off of silts and muds from the land. As well as clogging up intertidal shores, the resulting increases in water turbidity will ultimately reduce the growth of seaweeds.

A dramatic clogging up of the intertidal zone is witnessed following an oil spill. Seaweeds are not only susceptible to damage from oil washed ashore, but also to the vast quantities of detergents and other agents used to try to disperse the oil. Seaweeds do survive, but the species diversity of an oil-smothered shore will be greatly reduced, potentially taking many years to recover.

Seaweeds are a resilient group of marine organisms ideally suited to life in the harsh zone between the ocean and the land. They are key to the rich ecosystems that are typical of coastal waters and that must be conserved. Like every other resource we take from the sea, it is important that the exploitation of seaweed is correctly managed in a sustainable way to maintain natural indigenous populations. Mismanagement and unchecked coastal development will end in disaster. Coastal zone managers, policy makers and governments around the world need to be vigilant that we don't lose our seaweed resources.

Seaweed genera and species[*]

Rhodophyta (Red)

Ahnfeltiopsis concinna
Ahnfeltiopsis linearis
Bangia atropurpurea
Bostrychia
Caloglossa
Catenella
Ceramium
Chondrus
Chondrus crispus
Corallina
Delesseria
Gelidium
Gigartina
Gracilaria
Halosaccion glandiforme
Hypnea
Iridaea cordata
Jania
Jania rubens
Laurencia
Lithophyllum
Lithothamnion
Lithothamnion corallioides
Mastocarpus stellatus
Palmaria decipiens
Palmaria palmata
Peyssonnelia squamaria
Phymatolithon calcareum
Polysiphonia lanosa
Porolithon
Porphyra
Porphyra tenera
Pterocladia

Chlorophyta (Green)

Acetabularia
Acetabularia acetabulum
Caulerpa
Caulerpa taxifolia
Chaetomorpha
Chaetomorpha coliformis
Cladophora
Cladophora aegagrophila
Cladophora glomerata
Cladophora rupestris
Codium
Codium bursa
Enteromorpha intestinalis
Halimeda
Halimeda copiosa
Penicillus
Rhipocephalus phoenix
Udotea
Ulva
Ventricaria ventricosa

Heterokontophyta (class Phaeophyceae) (Brown)

Adenocystis
Alaria
Ascophyllum
Ascophyllum nodosum
Chorda
Chorda filum
Desmarestia
Dictyota
Durvillaea
Ecklonia
Ecklonia maxima
Ectocarpus
Ectocarpus siliculosus
Egregia laevigata
Elachista fucicola
Elachista scutulata
Fucus
Fucus spiralis
Fucus serratus
Fucus vesiculosus
Halidrys siliquosa
Himanthalia elongata
Hormosira banksii
Laminaria
Laminaria digitata
Laminaria hyperborea
Laminaria saccharina
Laminaria sinclairii
Laminaria solidungula
Leathesia difformis
Macrocystis
Macrocystis pyrifera
Nereocystis
Nereocystis luetkeana
Padina
Padina boergesenii
Pelagophycus porra
Pelvetia
Pelvetia canaliculata
Pilaiella littoralis
Postelsia
Saccorhiza polyschides
Sargassum
Sargassum muticum
Sphacelaria
Turbinaria
Undaria pinnatifida

* mentioned in this book

Glossary

Agar: a gelling agent extracted from the cell walls of some red seaweeds.

Alginate: salts extracted from brown seaweeds. Used extensively as gelling agents and various other industrial applications.

Alternation of generations: a type of life cycle in which there are two multicellular stages that can be distinguished by the type of reproduction and sometimes by structural features.

Antheridium (antheridia): a cell that divides to make male gametes.

Asexual reproduction: reproduction without the fusion of gametes.

Calcification: depositing calcium carbonate in the tissues of a seaweed.

Carotenoid: a fat-soluable yellow, orange or red pigment.

Carpospore: the spore released from a carposporangium in red algae.

Carposporangium (carposporangia): carpospore-producing structure.

Carrageenan: a gelling agent extracted from some red seaweeds.

Cell wall: a rigid, mutilayered structure surrounding the cell. It is made up of layers of large carbohydrates, lipids and proteins.

Chlorophyll: a fat-soluble green pigment. Various chlorophylls exist, but all seaweeds contain chlorophyll a.

Circadian rhythm: cyclic changes that are controlled by external factors on a 24-hour cycle.

Circannual rhythm: cyclic changes that are controlled by external factors on an annual cycle.

Conceptacle: cavity containing the reproductive cells of some seaweeds.

Conchocelis phase: a filamentous life cycle phase of some red seaweeds such as *Porphyra*. Found on shells and calcareous material.

Conchospores: spores produced during the **conchocelis phase** (e.g. *Porphyra*) of some red seaweed life cycles.

Cortex: a layer of cells lying between the epidermis of a seaweed and the inner medulla.

Cyanobacteria: photosynthetic bacteria that used to be called blue green algae. They are mostly unicellular, although they often grow in chains or colonies that make them large enough to see.

Diatoms: a group of microalgae characterised by hard external outer walls made from silicate (similar to glass).

Dioecious: male and female gametes are produced on separate individuals.

Diploid: having two sets of chromosomes in the same cell.

Diurnal: daily.

DOM: dissolved organic matter. A mixture of organic materials such as amino acids, carbohydrates and fats that are dissolved in seawater.

Ectocarpene: Pheromone (see below) released by the brown seaweeds *Ectocarpus, Sphacelaria* and *Adenocystis*.

Endemic: occurring only in a specific region or area.

Epiphyte: an organism that grows on the surface of a plant or seaweed.

Eutrophication: the addition of nutrients such as nitrogen and phosphorus to water bodies to make them nutrient rich.

Eutrophic: water bodies containing high levels of nutrients such as nitrogen and phosphorus. This can occur naturally, or as a result of eutrophication (see above).

Fronds: leaf-like blades of seaweeds.

Gamete: a cell (such as a sperm or egg) capable of fusing with another to form a zygote.

Gametophyte: the multicellular gamete-producing phase in the life cycle of seaweeds with alternation of generations (see above).

Haploid: having one set of chromosomes in each cell.

Herbivore: an organism that eats plant or algal material.

Holdfast: a structure that attaches seaweeds to a substratum.

Intertidal: a region generally covered by high tides and uncovered at low tides. Synonym for the littoral zone.

Littoral zone: a region generally covered by high tides, and uncovered at low tides. Synonym for the intertidal zone.

Medulla: region of cells in the interior of fleshy brown seaweeds.

Meiosis: nuclear division in which the chromosome number is halved.

Meristem: dividing tissue that adds new cells in a seaweed thallus. A region of growth.

Meristoderm: a surface layer of cells or epidermis on the surface of some seaweeds, especially brown species.

Macroalga (macroalgae): an alga large enough to be seen with the naked eye, although in some species details may be only seen using a good hand lens or even a microscope.

Microalga (microalgae): microscopic algae ranging in size from 20 to 200 microns.

Mitosis: nuclear division that results in both daughter nuclei being genetically identical to the parent nucleus.

Monoecious: both male and female gametes are produced on the same thallus.

Oogonium (oogonia): a structure producing female gametes (eggs).

Osmosis: the movement of pure water across a semi-permeable membrane from a region with low dissolved ions to one with a higher concentration of dissolved ions.

Paraphysis (paraphyses): a non-reproductive filament that occurs in sporangia or conceptacles of seaweeds.

Pheromone: a substance produced to invoke a behavioural response in another individual.

Photosynthesis: the transformation of light energy into chemical energy by cells. In the process carbon dioxide and water are transformed into sugars and oxygen.

Phycocyanin: a blue, water-soluble pigment found in red seaweeds.

Phycoerythrin: a red, water-soluble pigment found in red seaweeds.

Phytoplankton: floating microscopic microalgae.

Plankton: floating algae and small animals that cannot swim against a current, thereby floating where water currents take them.

Pneumatocyst: a bulbous gas filled structure that is used for buoyancy.

Polysaccharide: a large carbohydrate polymer made from simple sugar molecules.

Productivity: increase in seaweed biomass per unit of time in a given area.

Receptacle: region of a seaweed thallus in which reproductive structures develop.

Rhizoid: a filament that grows from the base of a seaweed thallus that helps seaweeds attach themselves.

Salinity: number of grams of dissolved salts in 1000 grams of seawater.

Sieve plates: porous end walls of cells making up sieve tubes (see next).

Sieve tubes: series of cells with porous end walls through which sugars are transported in some large brown seaweeds.

Sporangium (sporangia): a cell that divides to form spores. (See unilocular sporangium below.)

Spore: an asexual reproductive cell produced by sporangia.

Sporophyte: the multicellular spore-producing phase in the life cycle of seaweeds with alternation of generations (see above).

Stipe: part of a seaweed that joins the holdfast to the blade.

Sublittoral zone: parts of the shore below the lowest low tide mark.

Supralittoral zone: the shore above the uppermost high tide mark. May still be splashed by waves.

Thallus (thalli): the entire body of a seaweed.

Trumpet cells: elongated cells, wider at the ends than in the middle, found in the medulla tissues of some brown seaweeds.

Unilocular sporangium (sporangia): a sporangium (see above) that is a single cell that divides to produce spores. These contrast with plurilocular sporangia that are many chambered sporangia in which each chamber produces a spore.

Zooplankton: animals in the plankton.

Zygote: the product of two gametes fusing.

Index

Further Information

Recommended reading

Algae, L. Graham and L. Wilcox. Prentice Hall, 2000.

Algae and Human Affairs, C. Lembi and R.J. Waaland. Cambridge University Press, 1989.

Algae – An Introduction to Phycology, C. van den Hoek, D. Mann and H.M. Jahns. Cambridge University Press, 1996.

Killer Algae, A. Meinesz. The University of Chicago Press, 1999.

Marine Plants of Australia, J. M. Huisman. University of Western Australia Press, 2000.

Phycology, R.E. Lee. Cambridge University Press, 1999.

Seaweed – A User's Guide, S, Surey-Gent and G, Morris. Whittet Books, 2000.

Seaweed Resources in Europe – Uses and Potential, M.D. Guiry and G. Blunden. John Wiley & Sons, 1991.

Seaweeds: Ecology and Physiology, C. S. Lobban and P.J. Harrison. Cambridge University Press, 1997.

Seaweeds of the British Isles, Vols. 1-3 (in parts). Intercept Ltd.

Simply Seaweed, L. Ellis. Grub Street, 1998.

Academic journals reporting the latest algal discoveries provide a valuable resource of detailed information about seaweeds: *Botanica Marina* (Walter de Gruyter), *The European Journal of Phycology* (Cambridge University Press); *The Journal of Phycology* (Blackwell Science Ltd); *The Journal of Applied Phycology* (Kluwer), *Phycologia*. Several of these journals are from societies and organisations active in all areas of phycological research. A small selection is listed here:

Australasian Society for Phycology & Aquatic Botany
(http://www.possum.murdoch.edu.au/~cowan/)

British Phycological Society
(http://www.brphycsoc.org)

International Phycological Society
(http://www.intphycsoc.org)

The International Seaweed Association
(http://www.seaweed.ie/isa/)

Phycological Society of America
(http://www.psaalgae.org)

Recommended websites

NB. Website addresses are subject to change.

http://www.botany.uwc.ac.za/algae/
[A good introduction to the world of algae.]

http://www.marlin.ac.uk
[A comprehensive, easy to use source of information about marine habitats, communities and species around Britain and Ireland.]

http://www.mbayaq.org
[The Monterey Aquarium's website.]

http://www.nmnh.si.edu/botany/projects/algae/
[The algal group of the National Museum of Natural History, Smithsonian Institution. It contains a wealth of information, including extensive algal web links.]

http://www.seaweed.ie
[The best starting point for anyone interested in learning about all aspects of seaweed. Linked to this, Algal base (http://www.algaebase.com) is a dynamic, searchable database that stores information on more than 20,000 seaweed and seagrass species.]

http://www.tolweb.org
[A good site for the latest developments in our understanding of the classification and evolutionary relationships of seaweeds.]

http://www.weedseen.co.uk
[A site aiming to create an interactive colour photographic guide to a wide selection of marine algae of the British Isles.]

Picture credits

Front and back cover, title page, p.5 © D J Roberts; p.6 © J D George; pp.7 left and right, 8, 9, 10, 12 © D J Roberts; p.13 top left, middle left, bottom right © J D George; p.13 top right © C Wiencke; p.13 middle right, bottom left © D J Roberts; p.14 © D N Thomas; pp.15, 16 left © C Wiencke; p.16 right © G Russell; p.17 © D John; p.18 top G Russell; p.18 bottom © D J Roberts; p.19 © G Russell; p.20 left and right © J D George; p.21 © D J Roberts; p.22 Mike Eaton/© NHM; p.23 top left and right © G Russell; p.23 bottom Mike Eaton/© NHM; p.24 top left © D J Roberts; p.24 top right, bottom left and right © G Russell; p.25 top Mike Eaton/© NHM; p.25 bottom left © G Russell; p.25 bottom right © D N Thomas; p.26 top © M Guiry; p.26 bottom © D J Roberts; pp.27, 28 left and right © G Russell; p.29 top © D J Roberts; p.29 bottom © J D George; p.30 top © B Sanderson; p.30 bottom, p.32 © D J Roberts; pp.33, 34 © D N Thomas; p.35 top © D John; p.35 bottom © G Russell; p.36 © D J Roberts; p.37 © D Thomas; p.38 © D J Roberts; pp.39, 40 © G Russell; p.41 © D J Roberts; p.42 © G E Fogg; p.43 © D N Thomas; p.44 top © D J Roberts; p.44 bottom © D N Thomas; p.45 © D J Roberts; p.46 top © D N Thomas; p.46 bottom © A R Polanshek; p.47 © G Russell; p.48 © D N Thomas; p.49 top © G S Dieckmann; p.49 bottom © B Sanderson; p.50 © N Kamenos; p.51 © B Sanderson; p.52 © D N Thomas; p.53 © D J Roberts; p.54 © G S Dieckmann; p.55 top © J D George; p.55 bottom © D John; p.56 © David Shale/BBC Natural History Unit; p.57 © C Wiencke; p.58 © Jeff Foott/BBC Natural History Unit; pp.59, 60 © J D George; p.61 © G Russell; p.62 © D J Roberts; p.63 top and bottom © G Russell; p.64 top © D J Roberts; p.64 bottom © J Turner; p.65 © J D George; p.66 © D John; p.67 Hawkins and Hartnoll; p.68 © G Russell; p.69 © D N Thomas; p.70 top © G Russell; pp.70 bottom, 71, 72 © D N Thomas; p.73 © J Turner; p.74 © J D George; pp.75, 76 top © J Turner; pp.76 bottom, 77 © J D George; p.78 © J Turner; p.79 © D J Roberts; pp.80, 81 © J D George; p.82 © Alexandre Meinesz, University of Nice-Sophia Antipolis; p.84 © C Wiencke; pp.85, 86 © M Guiry; p.87 © G E Fogg; p.88 © B Sanderson

Author's acknowledgements

I am grateful to David Roberts, George Russell, Cornelia Thomas, Tony Fogg, Celia Coyne, Christian Wiencke, John Turner and Peter Williams for their help in the realisation of this project. This work was produced as a result of opportunities offered by the University of Wales, Bangor and the Hanse Institute for Advanced Study, Germany.